FOR "ANDY"

Andy— thanks so much for spawning another satisfied ologist David

To a dear friend and scientist whose fascination with BIOLOGY included the "BIG PICTURE" as well as the ultrastructure and its secrets, many of which Andy revealed.

Thanks for all the wonderful years! Betty

Andy— Thanks for providing the opportunities. Best wishes

Love

Very best wishes

Best wishes from Don Fawcett

With all of our best wishes, Frank & Kathy & kids

Andy— Thanks for all the sharing & learning over the many years. Howard

Andy— wish you the best in all the years to come

I count on your teaching forever Eli

Best wishes Pete

All the best, Andy! Eddington

Andy! You're still no big deal! Bruce

Congratulations Andy, Stan

THIS IS BIOLOGY

THIS IS BIOLOGY

The Science of the Living World

ERNST MAYR

THE BELKNAP PRESS OF
HARVARD UNIVERSITY PRESS
Cambridge, Massachusetts
London, England

Library of Congress Cataloging-in-Publication Data

Mayr, Ernst, 1904–

This is biology : the science of the living world / Ernst Mayr.

p. cm.

Includes bibliographical references and index.

ISBN 0-674-88468-X

1. Biology. I. Title.

QH307.2.M39 1997

574—dc20   96-42192

Designed by Gwen Frankfeldt

In memory of my mother,
Helene Pusinelli Mayr,
to whom I owe
so much

Contents

Preface

Some years ago the then-President of France, Valéry Giscard d'Estaing, declared the twentieth century "the century of biology." If this is perhaps not entirely accurate for the century as a whole, it has certainly been true for the second half. Today, biology is a thriving field of inquiry. We have witnessed unprecedented breakthroughs in genetics, cellular biology, and neuroscience, as well as spectacular advances in evolutionary biology, physical anthropology, and ecology. A whole industry has grown out of research in molecular biology; the results are readily apparent in fields as diverse as medicine, agriculture, animal breeding, and human nutrition, to name only a few.

The prospects for biology have not always been so bright. From the Scientific Revolution in the seventeenth century until well after World War II, science for most people meant the "exact" sciences—physics, chemistry, mechanics, astronomy—all of which relied heavily on mathematics and emphasized the role of universal laws. During this time physics was considered the exemplar of science. By comparison, the study of the living world was considered an inferior endeavor. Even today most people continue to hold profound misconceptions about the life sciences. For example, a failure to understand biology is frequently displayed in the media, whether the topic is the teaching of evolution, the measurement of intelligence, the possibility of detecting extraterrestrial life, the extinction of species, or the risks of smoking.

More regrettable, many biologists themselves have an obsolete notion of the life sciences. Modern biologists tend to be extreme specialists. They may know all about particular bird species, sex hormones, parental behavior, neuroanatomy, or the molecular structure of genes, yet they often are uninformed about developments outside their field of expertise. Rarely do biologists have the time to stand back from the advances in their own specialty and look at the life sciences as a whole. Geneticists, embryologists, taxonomists, and ecologists all consider themselves to be biologists, but most of them have little appreciation of what these various specialties have in common and how they differ fundamentally from the physical sciences. To shed some light on these issues is a major purpose of this book.

I have been a naturalist almost since I learned to walk, and my love of plants and animals led me to approach the living world holistically. Fortunately, the teaching of biology at the German high school I attended around 1920 centered on the whole organism and its interactions with the animate and inanimate environment. We would now say that the focus was on life history, behavior, and ecology. Physics and chemistry, both of which I also studied in high school, were something entirely different, and had little to do with living plants and animals.

During the years when I was a medical student, I was far too excited about medicine, and far too busy, to pay any attention to such basic questions as "What is biology?" and "What makes biology a science?" Indeed, there was not any subject taught at that time—at least not in the German universities—which was designated "biology." What we would now call biology was taught in departments of zoology and botany, both of which strongly emphasized the study of structural types and their phylogeny. To be sure, courses were also given in physiology, genetics, and other more or less experimental disciplines, but there was little integration of the subjects, and the conceptual framework of the experimentalists was largely incompatible with that of the zoologists and botanists, whose work was based in natural history.

After switching my studies from medicine to zoology (particularly birds) following the completion of my preclinical examinations, I took courses in philosophy at the University of Berlin. But to my disappointment, they built no bridges between the subject matter of the biological sciences and that of philosophy. Yet in the 1920s and 30s a discipline was developing that would eventually be designated "philosophy of science." In the 1950s, when I became acquainted with the teachings of this field, I was again bitterly disappointed. This was no philosophy of science; this was a philosophy of logic, mathematics, and the physical sciences. It had almost nothing to do with the concerns of biologists. Around this time I sat down and made a list of the major generalizations of evolutionary biology stated in books and published papers—a few of which, by this time, I had contributed myself—and found that not a single one of them was adequately covered in the philosophical literature; most of them were not even mentioned.

Still, at this point I had no plans to make a contribution to the history and philosophy of science. My various essays on these topics were the result of invitations to conferences and symposia, which forced me to lay aside temporarily my researches in evolutionary theory and systematics. My only intention was to point out how very different biology was in certain respects from physics. For example, in 1960 I was invited by Daniel Lerner of the Massachusetts Institute of Technology to participate in a lecture series dealing with cause and effect. I had been interested in the problem of biological causation since my Serin finch paper in 1926 and another paper on the origin of bird migration in 1930. Therefore I welcomed this opportunity to sort through my thoughts on this subject. I had long been aware of a categorical difference between the inanimate and the living world. Both worlds obey the universal laws discovered and analyzed by the physical sciences, but living organisms obey also a second set of causes, the instructions from the genetic program. This second type of causation is nonexistent in the inanimate world. Of course, I was not the first biologist to discover the duality of causation in organisms, but my 1961 published paper from the lecture series was the first to provide a detailed analysis of the subject.

In truth, my various essays about the differences between the life sciences and the physical sciences were directed not so much at

philosophers and physicists as at my fellow biologists, who had unwittingly adopted many physicalist concepts in their writings. For
example, the claim that every attribute of complex living systems can
be explained through a study of the lowest components (molecules,
genes, or whatever) struck me as absurd. Living organisms form a
hierarchy of ever more complex systems, from molecules, cells, and
tissues, through whole organisms, populations, and species. In each
higher system, characteristics emerge that could not have been predicted from a knowledge of the components.

At first I thought that this phenomenon of emergence, as it is now
called, was restricted to the living world; and indeed, in a lecture I
gave in the early 1950s in Copenhagen, I made the claim that emergence was one of the diagnostic features of the organic world. The
whole concept of emergence was at that time considered to be rather
metaphysical. When the physicist Niels Bohr, who was in the audience,
stood up to speak during the discussion period, I was fully prepared
for an annihilating refutation. However, much to my surprise, he did
not at all object to the concept of emergence but only to my notion
that it provided a demarcation between the physical and the biological
sciences. Citing the case of water, whose "aquosity" could not be
predicted from the characteristics of its two components, hydrogen
and oxygen, Bohr stated that emergence is rampant in the inanimate
world.

In addition to reductionism, another particularly objectionable bête
noire for me was typological thinking, later baptized "essentialism" by
the philosopher Karl Popper. It consisted of classifying the variation
of nature into fixed types (classes), invariant and sharply demarcated
against other such types. This concept, going back to Plato and Pythagorean geometry, was singularly unsuited to evolutionary and
population biology, where one finds not classes but aggregates of
unique individuals, that is, populations. Explaining variable phenomena in living nature in terms of populations—so-called population
thinking—seems to be difficult for those accustomed to physicalist
thinking. I repeatedly argued this problem with the physicist Wolfgang
Pauli, who was most anxious to understand what we biologists had
in mind. He finally came close to understanding it when I suggested

to him to think of a gas consisting of only 100 molecules, each differing from the others in direction and speed of movement. He called it an "individual gas."

Biology has also been misunderstood by many of those attempting to construct a history of science. When Thomas Kuhn's *Structure of Scientific Revolutions* was published in 1962, I was puzzled as to why it should have caused such a commotion. To be sure, Kuhn had refuted some of the most unrealistic theses of the traditional philosophy of science and had called attention to the importance of historical factors. But what he offered as a replacement seemed to me just as unrealistic. Where in the history of biology were the cataclysmic revolutions and where were the long periods of normal science postulated by Kuhn's theory? From what I knew of the history of biology, they did not exist. No doubt Darwin's *On the Origin of Species,* published in 1859, was revolutionary, but ideas about evolution had been in the air for a century. Moreover, Darwin's theory of natural selection—the key mechanism in evolutionary adaptation—was not fully accepted until almost a century after its publication. Throughout this time there were minor revolutions but never any period of "normal" science. Whether or not Kuhn's thesis was valid for the physical sciences, it did not fit biology. Historians coming from a background in physics seemed not to grasp what had happened in the study of living organisms over three centuries.

More and more clearly I began to see that biology was a quite different kind of science from the physical sciences; it differed fundamentally in its subject matter, its history, its methods, and its philosophy. While all biological processes are compatible with the laws of physics and chemistry, living organisms could not be reduced to these physicochemical laws, and the physical sciences could not address many aspects of nature that were unique to the living world. The classical physical sciences, on which the classical philosophy of science was based, were dominated by a set of ideas inappropriate to the study of organisms: these included essentialism, determinism, universalism, and reductionism. Biology, properly understood, comprises population thinking, probability, chance, pluralism, emergence, and historical narratives. What was needed was a new philosophy of science that could incorporate the approaches of all sciences, including physics and biology.

When planning this volume, I had in mind a more modest task, however. I wanted to write a "life history" of biology that would introduce the reader to the importance and richness of biology as a whole, while helping the individual biologist approach a problem that is steadily becoming more formidable: the information explosion. New workers annually join those already in the field and add to the avalanche of new publications. Practically all biologists I have ever talked with have complained to me that they no longer have time to keep up with the literature in their own specialty, much less adjacent disciplines. And yet often it is feedback from outside one's narrow domain that is decisive for a conceptual advance. New directions for research frequently come into view when one steps back from one's own field and sees it as part of a larger endeavor to explain the living world, in all its wonderful diversity. I hope that this book will provide a conceptual framework from which working biologists can attain this broader perspective on their specific research agenda.

Nowhere is the information explosion more apparent than in molecular biology. A detailed discussion of this field is absent from this volume not because I think molecular biology is less important than other parts of biology but for exactly the opposite reason. Whether we deal with physiology, development, genetics, neurobiology, or behavior, molecular processes are ultimately responsible for whatever happens, and every day researchers are making fresh discoveries in all these domains. In Chapters 8 and 9 I have highlighted some of the major generalizations ("laws") discovered by molecular biologists. Still, it strikes me that while we have identified many trees, we have not yet seen the forest. Others may disagree; in any case, a comprehensive overview of molecular biology requires a competence I do not have.

The same can be said for another exceedingly important discipline, the biology of mental processes. We are still in a stage of local exploration, and I simply do not command the required knowledge of neurobiology and psychology to attempt a broad analysis. A final area not covered in great detail in this volume is genetics. The genetic program plays a decisive role in every aspect of an organism's life: its structure, development, functions, and activities. Since the rise of molecular biology, the emphasis in genetics has shifted to developmental genetics, which has become virtually a branch of molecular biology, and for this reason I have not attempted to survey this field. However, I hope that my treatment of biology as a whole will be helpful in an eventual "life history" of these and other critical branches of biology that were not the direct focus of this volume.

If biologists, physical scientists, philosophers, historians, and others with a professional interest in the life sciences discover useful insights in the chapters that follow, this book will have accomplished one of its primary goals. But every educated person should have an understanding of basic biological concepts—evolution, biodiversity, competition, extinction, adaptation, natural selection, reproduction, development, and a host of others that are discussed in this book. Overpopulation, the destruction of the environment, and the malaise of the inner cities cannot be solved by technological advances, nor by literature or history, but ultimately only by measures that are based on an understanding of the biological roots of these problems. To "know thyself," as the ancient Greeks commanded us, entails first and foremost knowing our biological origins. To help readers gain a better understanding of our place in the living world, and of our responsibility to the rest of nature, is the major objective of this book.

Cambridge, Massachusetts
September 1996

THIS IS BIOLOGY

CHAPTER ONE

What Is the Meaning of "Life"?

Primitive humans lived close to nature. Every day they were occupied with animals and plants, as gatherers, hunters, or herdsmen. And death—of infants and elders, women in childbirth, men in strife—was forever present. Surely our earliest ancestors must have wrestled with the eternal question, "What is life?"

Perhaps, at first, no clear distinction was made between life in a living organism and a spirit in a nonliving natural object. Most primitive people believed that a spirit might reside in a mountain or a spring as well as in a tree, an animal, or a person. This animistic view of nature eventually waned, but the belief that "something" in a living creature distinguished it from inanimate matter and departed from the body at the moment of death held strong. In ancient Greece this something in humans was referred to as "breath." Later, particularly in the Christian religion, it was called the soul.

By the time of Descartes and the Scientific Revolution, animals (along with mountains, rivers, and trees) had lost their claim to a soul. But a dualistic split between body and soul in human beings continued to be almost universally accepted and is even today still believed by many people. Death was a particularly puzzling problem for a dualist. Why should this soul suddenly either die or leave the body? If the soul left the body, did it go somewhere, such as to some nirvana or heaven? Not until Charles Darwin developed his theory of evolution through natural selection was a scientific, rational explanation for death possible. August Weismann, a follower of Darwin at the end of the nineteenth century, was the first author to explain that a rapid sequence of generations provides the number of new genotypes required to cope permanently with a changing environment. His essay on death and dying was the beginning of a new era in our understanding of the meaning of death.

When biologists and philosophers speak of "life," however, they usually are not referring to life (that is, living) as contrasted with death but rather to life as contrasted with the lifelessness of an inanimate object. To elucidate the nature of this entity called "life" has been one of the major objectives of biology. The problem here is that "life" suggests some "thing"—a substance or force—and for centuries philosophers and biologists have tried to identify this life substance or vital force, to no avail. In reality, the noun "life" is merely a reification of the process of living. It does not exist as an independent entity.[1] One can deal with the process of living scientifically, something one cannot do with the abstraction "life." One can describe, even attempt to define, what living is; one can define what a living organism is; and one can attempt to make a demarcation between living and nonliving. Indeed, one can even attempt to explain how living, as a process, can be the product of molecules that themselves are not living.[2]

What life is, and how one should explain living processes, has been a subject of heated controversy since the sixteenth century. In brief, the situation was this: There was always a camp claiming that living organisms were not really different at all from inanimate matter; sometimes these people were called mechanists, later physicalists. And there was always an opposing camp—called vitalists—claiming instead that living organisms had properties that could not be found in inert matter and that therefore biological theories and concepts could not be reduced to the laws of physics and chemistry. In some periods and at certain intellectual centers the physicalists seemed to be victorious, and in other times and places the vitalists seemed to have achieved the upper hand. In this century it has become clear that both camps were partly right and partly wrong.

The physicalists had been right in insisting that there is no metaphysical life component and that at the molecular level life can be explained according to the principles of physics and chemistry. At the same time, the vitalists had been right in asserting that, nevertheless, living organisms are not the same as inert matter but have numerous autonomous characteristics, particularly their historically acquired genetic programs, that are unknown in inanimate matter. Organisms are many-level ordered systems, quite unlike anything found in the inanimate world. The philosophy that eventually incorporated the best principles from both physicalism and vitalism (after discarding the excesses) became known as organicism, and this is the paradigm that is dominant today.

The Physicalists

Early beginnings of a natural (as opposed to supernatural) explanation of the world were made in the philosophies of various Greek thinkers, including Plato, Aristotle, Epicurus, and many others. These promising beginnings, however, were largely forgotten in later centuries. The Middle Ages were dominated by a strict adherence to the teachings of the Scriptures, which attributed everything in nature to God and His laws. But medieval thinking, particularly in folklore, was also characterized by a belief in all sorts of occult forces. Eventually this animistic, magical thinking was reduced, if not eliminated, by a new way of looking at the world that was aptly called "the mechanization of the world picture" (Maier 1938).[3]

The influences leading up to the mechanization of the world picture were manifold. They included not only the Greek philosophers, transmitted to the Western world by the Arabs along with rediscovered original writings, but also technological developments in late medieval and early Renaissance times. There was great fascination with clocks and other automata—and indeed with almost any kind of machine. This eventually culminated in Descartes's claim that all organisms except humans were nothing *but* machines.

Descartes (1596–1650) became the spokesman for the Scientific Revolution, which, with its craving for precision and objectivity, could not accept vague ideas, immersed in metaphysics and the supernatural, such as souls of animals and plants. By restricting the possession of a soul to humans and by declaring animals to be nothing but automata, Descartes cut the Gordian knot, so to speak. With the mechanization of the animal soul, Descartes completed the mechanization of the world picture.[4]

It is a little difficult to understand why the machine concept of organisms could have had such long-lasting popularity. After all, no machine has ever built itself, replicated itself, programmed itself, or been able to procure its own energy. The similarity between an organism and a machine is exceedingly superficial. Yet the concept did not die out completely until well into this century.

The success of Galileo, Kepler, and Newton in using mathematics to reinforce their explanations of the cosmos also contributed to the mechanization of the world picture. Galileo (1623) succinctly captured the prestige of mathematics in the Renaissance when he said that the book of nature "cannot be understood unless one first learns to comprehend the language and read the letters in which it is composed. It is written in the language of mathematics, and its characters are triangles, circles, and other geometric figures without which it is humanly impossible to understand a single word of it; without these one wanders about in a dark labyrinth."

The rapid development of physics shortly thereafter carried the Scientific Revolution a step further, turning the more general mechanicism of the early period into a more specific physicalism, based on a set of concrete laws about the workings of both the heavens and the earth.[5]

The physicalist movement had the enormous merit of refuting much of the magical thinking that had generally characterized the preceding centuries. Its greatest achievement perhaps was providing a natural explanation of physical phenomena and eliminating much of the reliance on the supernatural that was previously accepted by virtually everybody. If mechanicism, and particularly its outgrowth into physicalism, went too far in some respects, this was inevitable for an energetic new movement. Yet because of its one-sidedness and its failure to explain any of the phenomena and processes particular to living organisms, physicalism induced a rebellion. This countermovement is usually described under the umbrella term vitalism.

From Galileo to modern times there has been a seesawing in biology between strictly mechanistic and more vitalistic explanations of life. Eventually, Cartesianism reached its culmination in the publication of de La Mettrie's *L'homme machine* (1749). Next followed a vigorous flowering of vitalism, particularly in France and in Germany, but further triumphs of physics and chemistry in the mid-nineteenth century inspired yet another physicalist resurgence in biology. It was largely confined to Germany, perhaps not surprisingly so, since nowhere else did biology flourish in the nineteenth century to the extent it did in Germany.

THE FLOWERING OF PHYSICALISM

The nineteenth-century physicalist movement arrived in two waves. The first one was a reaction to the quite moderate vitalism adopted by Johannes Müller (1801–1858), who in the 1830s switched from pure physiology to comparative anatomy, and of Justus von Liebig (1803–1873), well known for his incisive critiques which helped to bring the reign of inductivism to an end. It was set in motion by four former students of Müller—Hermann Helmholtz, Emil DuBois-Reymond, Ernst Brücke, and Matthias Schleiden. The second wave, which began around 1865, is identified with the names Carl Ludwig, Julius Sachs, and Jacques Loeb. Undeniably, these physicalists made important contributions to physiology. Helmholtz (along with Claude Bernard in France) deprived "animal heat" of its vitalistic connotation, and DuBois-Reymond dispelled much of the mystery of nerve physiology by offering a physical (electric) explanation of nerve activity. Schleiden advanced the fields of botany and cytology through his insistence that plants consist entirely of cells and that all the highly diverse structural elements of plants are cells or cell products. Helmholtz, DuBois-Reymond, and Ludwig were particularly outstanding in the invention of ever-more sophisticated instruments to record the precise measurements in which they were interested. This permitted them, among other achievements, to rule out the existence of a "vital force" by showing that work could be translated into heat without residue. Every history of physiology written since that time has documented these and other splendid accomplishments.

Yet, the underlying philosophy of this physicalist school was quite naive and could not help but provoke disdain among biologists with a background in natural history. In historical accounts of the many achievements of the physicalists, their naivete when it came to living processes has frequently been ignored. But one cannot understand the vitalists' passionate resistance to the claims of the physicalists unless one is acquainted with the actual explanatory statements the physicalists offered.

It is ironic that the physicalists attacked the vitalists for invoking an unanalyzed "vital force," and yet in their own explanations they used such equally unanalyzed factors as "energy" and "movements." The definitions of life and the descriptions of living processes formulated by the physicalists often consisted of utterly vacuous statements. For example, the physical chemist Wilhelm Ostwald defined a sea urchin as being, like any other piece of matter, "a spatially discrete cohesive sum of quantities of energy." For many physicalists, an unacceptable vitalistic statement became acceptable when vital force was replaced by the equally undefined term "energy." Wilhelm Roux (1895), whose work brought experimental embryology into full flower, stated that development is "the production of diversity owing to the unequal distribution of energy."

Even more fashionable than "energy" was the term "movement" to explain living processes, including developmental and adaptational ones. DuBois-Reymond (1872) wrote that the understanding of nature "consists in explaining all changes in the world as produced by the movement of atoms," that is, "by reducing natural processes to the mechanics of atoms . . . By showing that the changes in all natural bodies can be explained as a constant sum . . . of potential and kinetic energy, nothing in these changes remains to be further explained." His contemporaries did not notice that these assertions were only empty words, without substantial evidence and with precious little explanatory value.

A belief in the importance of the movement of atoms was held not only by the physicalists but even by some of their opponents. For Rudolf Kölliker (1886)—a Swiss cytologist who recognized that the chromosomes in the nucleus are involved in inheritance and that spermatozoa are cells—development was a strictly physical phenomenon controlled by differences in growth processes: "It is sufficient to postulate the occurrence in the nuclei of regular and typical movements controlled by the structure of the idioplasm."

As exemplified in statements by the botanist Karl Wilhelm von Nägeli (1884), another favorite explanation of the mechanists was to invoke "movements of the smallest parts" to explain "the mechanics of organic life."[6] The effect of a nucleus on the rest of the cell—the cytoplasm—was seen by E. Strasburger, a leading botanist of the time, as "a propagation of molecular movements . . . in a manner which might be compared to the transmission of a nervous impulse." Thus it did not involve the transport of material; this notion was, of course, entirely wrong. These physicalists never noticed that their statements about energy and movement did not really explain anything at all. Movements, unless directed, are random, like Brownian motion. Something has to give direction to these movements, and this is exactly what their vitalist opponents always emphasized.

The weakness of a purely physicalist interpretation was particularly obvious in explanations of fertilization. When F. Miescher (a student of His and Ludwig) discovered nucleic acid in 1869, he thought that the function of the spermatozoon was the purely mechanical one of getting cell division going; as a consequence of his physicalist bias, Miescher completely missed the significance of his own discovery. Jacques Loeb claimed that the really crucial agents in fertilization were not the nucleins of the spermatozoon but the ions. One is almost embarrassed when reading Loeb's statement that "Branchipus is a freshwater crustacean which, if raised in concentrated salt solution, becomes smaller and undergoes some other changes. In that case it is called Artemia." The sophistication of the physicalists in chemistry, particularly physical chemistry, was not matched by their biological knowledge. Even Sachs, who studied so diligently the effects of various extrinsic factors on growth and differentiation, never seems to have given any thought to the question why seedlings of different species of plants raised under identical conditions of light, water, and food would give rise to entirely different species.

Perhaps the most uncompromising mechanistic school in modern biology was that of Entwicklungsmechanik, founded in the 1880s by Wilhelm Roux. This school of embryology represented a rebellious reaction to the one-sidedness of the comparative embryologists, who were interested only in phylogenetic questions. Roux's associate, the embryologist Hans Driesch, was at first, if anything, even more mechanistic, but he eventually experienced a complete conversion from an extreme mechanist to an extreme vitalist. This happened when he separated a sea urchin embryo at the two-cell stage into two separate embryos of one cell each and observed that these two embryos did not develop into two half organisms, as his mechanistic theories demanded, but were able to compensate appropriately and develop into somewhat smaller but otherwise perfect larvae.

In due time, the vacuousness and even absurdity of these purely physicalistic explanations of life became apparent to most biologists, who, however, were usually satisfied to adopt the agnostic position that organisms and living processes simply could not be exhaustively explained by reductionist physicalism.

The Vitalists

The problem of explaining "life" was the concern of the vitalists from the Scientific Revolution until well into the nineteenth century; it did not really become the subject matter of scientific analysis until the rise of biology after the 1820s. Descartes and his followers had been unable to persuade most students of plants and animals that there were no essential differences between living organisms and inanimate matter. Yet after the rise of physicalism, these naturalists had to take a new look at the nature of life and attempted to advance *scientific* (rather than metaphysical or theological) arguments against Descartes's machine theory of organisms. This requirement led to the birth of the vitalistic school of biology.[7]

The reactions of the vitalists to physicalist explanations were diversified, since the physicalist paradigm itself was composite, not only in what it claimed (that living processes are mechanistic and can be reduced to the laws of physics and chemistry) but also in what it failed to take account of (the differences between living organisms and simple matter, the existence of adaptive but much more complex properties—Kant's Zweckmässigkeit—in animals and plants, and evolutionary explanations). Each of these claims and omissions was criticized by one or the other opponent of physicalism. Some vitalists focused on unexplained vital properties, others on the holistic nature of living creatures, still others on adaptedness or directedness (as in the development of the fertilized egg).

All these opposing arguments to the various aspects of physicalism have traditionally been lumped together as vitalism. In some sense, this is not altogether wrong, because all of the antiphysicalists defended the life-specific properties of living organisms. Yet the label vitalist conceals the heterogeneity of this group.[8] For instance, in Germany some biologists (which Lenoir calls teleomechanists) were willing to explain physiological processes mechanically but insisted that this failed to account for either adaptation or directed processes, such as the development of the fertilized egg. These legitimate questions were raised again and again by distinguished philosophers and biologists from 1790 until the end of the nineteenth century, but they had remarkably little effect on the writings of the leading physicalists such as Ludwig, Sachs, or Loeb.

Vitalism, from its emergence in the seventeenth century, was decidedly an antimovement. It was a rebellion against the mechanistic philosophy of the Scientific Revolution and against physicalism from Galileo to Newton. It passionately resisted the doctrine that the animal is nothing but a machine and that all manifestations of life can be exhaustively explained as matter in motion. But as decisive and convincing as the vitalists were in their rejection of the Cartesian model, they were equally indecisive and unconvincing in their own explanatory endeavors. There was great explanatory diversity but no cohesive theory.

Life, according to one group of vitalists, was connected either with a special substance (which they called protoplasm) not found in inanimate matter, or with a special state of matter (such as the colloidal state), which, it was claimed, the physicochemical sciences were not equipped to analyze. Another subset of vitalists held that there is a special vital force (sometimes called Lebenskraft, Entelechie, or élan vital) distinct from the forces physicists deal with. Some of those who accepted the existence of such a force were also teleologists who believed that life existed for some ultimate purpose. Other authors invoked psychological or mental forces (psychovitalism, psycho-Lamarckism) to account for aspects of living organisms that the physicalists had failed to explain.

Those who supported the existence of a vital force had highly diverse views of the nature of this force. From about the middle of the seventeenth century on, the vital agent was most frequently characterized as a fluid (not a liquid), in analogy to Newton's gravity and to caloric, phlogiston, and other "imponderable fluids." Gravity was invisible and so was the heat that flowed from a warm to a cold object; hence, it was not considered disturbing or unlikely that the vital fluid was also invisible, even though not necessarily something supernatural. For instance, the influential late eighteenth-century German naturalist J. F. Blumenbach (who wrote extensively on extinction, creation, catastrophes, mutability, and spontaneous generation) considered this vital fluid, though invisible, to be nevertheless very real and subject to scientific study, much as gravity was.[9] The concept of a vital fluid was eventually replaced by that of a vital force. Even such a reputable scientist as Johannes Müller accepted a vital force as indispensable for explaining the otherwise inexplicable manifestations of life.

In England, all the physiologists of the sixteenth, seventeenth, and eighteenth centuries had vitalistic ideas, and vitalism was still strong in the 1800–1840 period in the writings of J. Hunter, J. C. Prichard, and others. In France, where Cartesianism had been particularly powerful, it is not surprising that the vitalists' countermovement was equally vigorous. The outstanding representatives in France were the Montpellier school (a group of vitalistic physicians and physiologists) and the histologist F. X. Bichat. Even Claude Bernard, who studied such functional subjects as the nervous and digestive systems and considered himself an opponent of vitalism, actually supported a number of vitalistic notions. Furthermore, most Larmarckians were rather vitalistic in some of their thinking.

It was in Germany that vitalism had its most extensive flowering and reached its greatest diversity. Georg Ernst Stahl, a late seventeenth-century chemist and physician best known for his phlogiston theory of combustion, was the first great opponent of the mechanists. Perhaps he was more of an animist than a vitalist, but his ideas played a large role in the teaching of the Montpellier school.

The next impetus to the vitalistic movement in Germany was the preformation versus epigenesis controversy, which dominated developmental biology in the second half of the eighteenth century. Preformationists held that the parts of an adult exist in smaller form at the very beginning of development. The epigenesists held that the adult parts appear as products of development but are not present as parts in the beginning. In 1759, when the embryologist Caspar Friedrich Wolff refuted preformation and replaced it by epigenesis, he had to invoke some causal agent that would convert the completely unformed mass of the fertilized egg into the adult of a particular species. He called this agent the *vis essentialis.*

J. F. Blumenbach rejected the vague *vis essentialis* and proposed instead that a specific formative force, *nisus formativus,* plays a decisive role not only in the development of the embryo but also in growth, regeneration, and reproduction. He accepted still other forces, such as irritability and sensibility, as contributing to the maintenance of life. Blumenbach was quite pragmatic about these forces, considering them essentially as labels for observed processes of which he did not know the causes. They were black boxes for him, rather than metaphysical principles.

The branch of German philosophy called Naturphilosophie, advanced by F. W. J. Schelling and his followers early in the nineteenth century, was a decidedly metaphysical vitalism, but the practical philosophies of working biologists such as Wolff, Blumenbach, and eventually Müller were antiphysicalist rather than metaphysical. Müller has been maligned as an unscientific metaphysician, but the accusation is unfair. A collector of butterflies and plants from his boyhood on, he had acquired the naturalist's habit of looking at organisms holistically. This perception was lacking in his students, whose leanings were more toward mathematics and the physical sciences. Müller realized that the slogan "life is a movement of particles" was meaningless and without explanatory value, and his alternative concept of Lebenskraft (vital force), though a failure, was closer to the concept of a genetic program than the shallow physicalist explanations of his rebellious students.[10]

Many of the arguments put forth by the vitalists were intended to explain specific characteristics of organisms which today are explained by the genetic program. They advanced a number of perfectly valid refutations of the machine theory but, owing to the backward state of biological explanation available at that time, were unable to come up with the correct explanation of vital processes that were eventually found during the twentieth century. Consequently, most of the argumentation of the vitalists was negative. From the 1890s on Driesch argued, for example, that physicalism could not explain self-regulation in embryonic structures, regeneration and reproduction, and psychic phenomena, like memory and intelligence. Yet it is remarkable how often perfectly sensible sentences emerge in Driesch's writings whenever his word "Entelechie" is replaced by the phrase "genetic program." These vitalists not only knew that there was something missing in the mechanistic explanations but they also described in detail the nature of the phenomena and processes the mechanists were unable to explain.[11]

Given the many weaknesses and even contradictions in vitalist explanations, it may seem surprising how widely vitalism was adopted and how long it prevailed. One reason, as we have seen, is that at that time there was simply no other alternative to the reductionist machine theory of life, which, to many biologists, was clearly out of the question. Another reason is that vitalism was strongly supported by several other then-dominant ideologies, including the belief in a cosmic purpose (teleology or finalism). In Germany, Immanuel Kant had a strong influence on vitalism, particularly on the school of teleomechanism, an influence still evident in Driesch's writings. A close connection with finalism is evident in the writings of most vitalists.[12]

In part because of their teleological leanings, the vitalists strongly opposed Darwin's selectionism. Darwin's theory of evolution denied the existence of any cosmic teleology and substituted in its place a "mechanism" for evolutionary change—natural selection: "We see in Darwin's discovery of natural selection in the struggle for existence the most decisive proof for the exclusive validity of mechanically operating causations in the whole realm of biology, and we see in this the definitive demise of all teleological and vitalistic interpretations of organisms" (Haeckel 1866). Selectionism made vitalism superfluous in the realm of adaptation.

Driesch was a rabid anti-Darwinian, as were other vitalists, but his arguments against selection were consistently ridiculous and showed clearly that he did not in the least understand this theory. Darwinism, by supplying a mechanism for evolution while at the same time denying any finalistic or vitalistic view of life, became the foundation of a new paradigm to explain "life."

THE DECLINE OF VITALISM

When vitalism was first proposed and widely adopted, it seemed to provide a reasonable answer to the nagging question, "What is life?" Furthermore, at that time it was a legitimate theoretical alternative not just to the crude mechanicism of the Scientific Revolution but also to nineteenth-century physicalism. Vitalism seemingly explained the manifestations of life far more successfully than the simplistic machine theory of its opponents.

Yet considering how dominant vitalism was in biology and for how long a period it prevailed, it is surprising how rapidly and completely it collapsed. The last support of vitalism as a viable concept in biology disappeared about 1930. A considerable number of different factors contributed to its downfall.

First, vitalism was more and more often viewed as a metaphysical rather than a scientific concept. It was considered unscientific because the vitalists had no method to test it. By dogmatically asserting the existence of a vital force, the vitalists often impeded the pursuit of a constitutive reductionism that would elucidate the basic functions of living organisms.

Second, the belief that organisms were constructed of a special substance quite different from inanimate matter gradually lost support. That substance, it was believed through most of the nineteenth century, was protoplasm, the cellular material outside the nucleus.[13] Later it was called cytoplasm (a term introduced by Kölliker). Because proto-plasm seemed to have what was called "colloidal" properties, a flourishing branch of chemistry developed: colloidal chemistry. Biochem-istry, however, together with electron microscopy, eventually estab-lished the true composition of cytoplasm and elucidated the nature of its various components: cellular organelles, membranes, and mac-romolecules. It was found that there was no special substance "pro-toplasm," and the word and concept disappeared from the biological literature. The nature of the colloidal state was likewise explained biochemically, and colloidal chemistry ceased to exist. Thus all evidence for a separate category of living substance disappeared, and it became possible to explain the seemingly unique properties of living matter in terms of macromolecules and their organization. The macromole-cules, in turn, are composed of the same atoms and small molecules as inanimate matter. Wöhler's synthesis in the laboratory of the organic substance urea in 1828 was the first proof of the artificial conversion of inorganic compounds into an organic molecule.

Third, all of the vitalists' attempts to demonstrate the existence of a nonmaterial vital force ended in failure. Once physiological and developmental processes began to be explained in terms of physico-chemical processes at the cellular and molecular level, these explana-tions left no unexplained residue that would require a vitalistic inter-pretation. Vitalism simply became superfluous.

Fourth, new biological concepts to explain the phenomena that used to be cited as proof of vitalism were developed. Two advances in particular were crucial for this change. One was the rise of genetics, which ultimately led to the concept of the genetic program. This made it possible to explain all goal-directed living phenomena, at least in principle, as teleonomic processes controlled by genetic programs. Another seemingly teleological phenomenon to be newly interpreted was Kant's Zweckmässigkeit. This reinterpretation was achieved by the second advance, Darwinism. Natural selection made adaptedness pos-sible by making use of the abundant variability of living nature. Thus, two major ideological underpinnings of vitalism—teleology and an-tiselectionism—were destroyed. Genetics and Darwinism succeeded in providing valid interpretations of the phenomena claimed by the vitalists not to be explicable except by invoking a vital substance or force.

If one were to believe the writings of the physicalists, vitalism was nothing but an impediment to the growth of biology. Vitalism took the phenomena of life, so it was claimed, out of the realm of science and transferred them to the realm of metaphysics. This criticism is indeed justified for the writings of some of the more mystical vitalists, but it is not fair when raised against reputable scientists such as Blumenbach and, even more so, Müller, who specifically articulated all the aspects of life that were left unexplained by the physicalists. That the explanation Müller adopted was a failure does not diminish the merit of his having outlined the problems that still had to be solved.

There are many similar situations in the history of science where unsuitable explanatory schemes were adopted for a clearly visualized problem because the groundwork for the real explanation had not yet been laid. Kant's explanation of evolution by teleology is a famous example. It is probably justifiable to conclude that vitalism was a necessary movement to demonstrate the vacuity of a shallow physicalism in the explanation of life. Indeed, as François Jacob (1973) has rightly stated, the vitalists were largely responsible for the recognition of biology as an autonomous scientific discipline.

Before turning to the organicist paradigm which replaced both vitalism and physicalism, we might note in passing a rather peculiar twentieth-century phenomenon—the development of vitalistic beliefs among physicists. Niels Bohr was apparently the first to suggest that special laws not found in inanimate nature might operate in organisms. He thought of these laws as analogous to the laws of physics except for their being restricted to organisms. Erwin Schrödinger and other physicists supported similar ideas. Francis Crick (1966) devoted a whole book to refuting the vitalistic ideas of the physicists Walter Elsasser and Eugene Wigner. It is curious that a form of vitalism survived in the minds of some reputable physicists long after it had become extinct in the minds of reputable biologists.

A further irony, however, is that many biologists in the post-1925 period believed that the newly discovered principles of physics, such as the relativity theory, Bohr's complementarity principle, quantum mechanics, and Heisenberg's indeterminacy principle, would offer new insight into biological processes. In fact, so far as I can judge, none of these principles of physics applies to biology. In spite of Bohr's searching in biology for evidence of complementarity, and some desperate analogies to establish this, there really is no such thing in biology as that principle. The indeterminacy of Heisenberg is something quite different from any kind of indeterminacy encountered in biology.

Vitalism survived even longer in the writings of philosophers than in the writings of physicists. But so far as I know, there are no vitalists among the group of philosophers of biology who started publishing after 1965. Nor do I know of a single reputable living biologist who still supports straightforward vitalism. The few late twentieth-century biologists who had vitalistic leanings (A. Hardy, S. Wright, A. Portmann) are no longer alive.

The Organicists

By about 1920 vitalism seemed to be discredited. The physiologist J. S. Haldane (1931) stated quite rightly that "biologists have almost unanimously abandoned vitalism as an acknowledged belief." At the same time, he also said that a purely mechanistic interpretation cannot account for the coordination that is so characteristic of life. What particularly puzzled Haldane was the orderly sequence of events during development. After showing the invalidity of both the vitalistic and the mechanistic approaches, Haldane stated that "we must find a different theoretical basis of biology, based on the observation that all the phenomena concerned tend towards being so coordinated that they express what is normal for an adult organism."

The demise of vitalism, rather than leading to the victory of mechanicism, resulted in a new explanatory system. This new paradigm accepted that processes at the molecular level could be explained exhaustively by physicochemical mechanisms but that such mechanisms played an increasingly smaller, if not negligible, role at higher levels of integration. There they are supplemented or replaced by emerging characteristics of the organized systems. The unique characteristics of living organisms are not due to their composition but rather to their organization. This mode of thinking is now usually referred to as *organicism*. It stresses particularly the characteristics of highly complex ordered systems and the historical nature of the evolved genetic programs in organisms.

According to W. E. Ritter, who coined the term organicism in 1919,[14] "Wholes are so related to their parts that not only does the existence of the whole depend on the orderly cooperation and interdependence of its parts, but the whole exercises a measure of determinative control over its parts" (Ritter and Bailey 1928). J. C. Smuts (1926) explained his own holistic view of organisms as follows: "A whole according to the view here presented is not simple, but composite and consists of parts. Natural wholes, such as organisms, are . . . complex or composite, consisting of many parts in active relation and interaction of one kind or another, and the parts may be themselves lesser wholes, such as cells in an organism." His statements were later condensed by other biologists into the concise statement that "a whole is more than the sum of its parts."[15]

Since the 1920s, the terms holism and organicism have been used interchangeably. Perhaps, at first, holism was more frequently used, and the adjective "holistic" is still useful today. But holism is not a strictly biological term, since many inanimate systems are also holistic, as Niels Bohr has pointed out correctly. Therefore, in biology the more restricted term "organicism" is now used more frequently. It encompasses the recognition that the existence of a genetic program is an important feature of the new paradigm.

The objection of the organicists was not so much to the mechanistic aspects of physicalism as to its reductionism. The physicalists referred to their explanations as mechanistic explanations, which indeed they were, but what characterized them far more was that they were also reductionist explanations. For reductionists, the problem of explanation is in principle resolved as soon as the reduction to the smallest components has been accomplished. They claim that as soon as one has completed the inventory of these components and has determined the function of each of them, it should be an easy task to explain also everything observed at the higher levels of organization.

The organicists demonstrated that this claim is simply not true, because explanatory reductionism is quite unable to explain characteristics of organisms that emerge at higher levels of organization. Curiously, even most mechanists admitted the insufficiency of a purely reductionist explanation. The philosopher Ernest Nagel (1961), for instance, conceded "that there are large sectors of biological study in which physico-chemical explanations play no role at present, and that a number of outstanding biological theories have been successfully exploited which are not physico-chemical in character." Nagel tried to save reductionism by inserting the words "at present," but it was already rather evident that such purely biological concepts as territory, display, predator thwarting, and so on could never be reduced to the terms of chemistry and physics without entirely losing their biological meaning.[16]

The pioneers of holism (for example, E. S. Russell and J. S. Haldane) argued effectively against the reductionist approach and described convincingly how well a holistic approach fits the phenomena of behavior and development. But they failed to explain the actual nature of the holistic phenomena. They were unsuccessful when trying to explain the nature of "the whole" or the integration of parts into the whole. Ritter, Smuts, and other early proponents of holism were equally vague (and somewhat metaphysical) in their explanations. Indeed, some of Smuts's wordings had a rather teleological flavor.[17]

Alex Novikoff (1947), however, spelled out in considerable detail why an explanation of living organisms has to be holistic. "What are wholes on one level become parts on a higher one . . . both parts and wholes are material entities, and integration results from the interaction of parts as a consequence of their properties." Holism, since it rejects reduction, "does not regard living organisms as machines made of a multitude of discrete parts (physico-chemical units), removable like pistons of an engine and capable of description without regard to the system from which they are removed." Owing to the interaction of the parts, a description of the isolated parts fails to convey the properties of the system as a whole. It is the organization of these parts that controls the entire system.

There is an integration of the parts at each level, from the cell to tissues, organs, organ systems, and whole organisms. This integration is found at the biochemical level, at the developmental level, and in whole organisms at the behavioral level.[18] All holists agree that no system can be exhaustively explained by the properties of its isolated components. The basis of organicism is the fact that living beings have organization. They are not just piles of characters or molecules, because their function depends entirely on their organization, their mutual interrelations, interactions, and interdependencies.

EMERGENCE

It is now clear that two major pillars in the explanatory framework of modern biology were missing in all the early presentations of holism. One, the concept of the genetic program, was absent because it had not yet been developed. The other missing pillar was the concept of emergence—that in a structured system, new properties emerge at higher levels of integration which could not have been predicted from a knowledge of the lower-level components. This concept was absent because either it had not been thought of or it had been dismissed as unscientific and metaphysical. By eventually incorporating the concepts of the genetic program and of emergence, organicism became antireductionist and yet remained mechanistic.

Jacob (1973) describes emergence this way: "At each level, units of relatively well-defined size and almost identical structure associate to form a unit of the level above. Each of these units formed by the integration of sub-units may be given the general name 'integron'. An integron is formed by assembling integrons of the level below it; it takes part in the construction of the integron of the level above." Each integron has new characteristics and capacities not present at any lower level of integration; these can be said to have emerged.[19]

The concept of emergence first received prominence in Lloyd Morgan's book on emergent evolution (1923). Darwinians who adopted emergent evolution nevertheless had some misgivings about it because they were afraid that it was antigradualistic. Indeed, some early emergentists were also saltationists, particularly during the period of Mendelism; that is, they believed that evolution proceeded in large, discontinuous leaps, or saltations. These misgivings have now been overcome, because it is now understood that the population (or species), rather than the gene or the individual, is the unit of evolution; one can have different forms (phenetic discontinuities) within populations—by recombination of existing DNA—while a population as a whole must by necessity evolve gradually. A modern evolutionist would say that the formation of a more complex system, representing the emergence of a new higher level, is strictly a matter of genetic variation and selection. Integrons evolve through natural selection, and at every level they are adapted systems, because they contribute to the fitness of an individual. This in no way conflicts with the principles of Darwinism.

To sum up, organicism is best characterized by the dual belief in the importance of considering the organism as a whole, and at the same time the firm conviction that this wholeness is not to be considered something mysteriously closed to analysis but that it should be studied and analyzed by choosing the right level of analysis. The organicist does not reject analysis but insists that analysis should be continued downward only to the lowest level at which this approach yields relevant new information and new insights. Every system, every integron, loses some of its characteristics when taken apart, and many of the important interactions of components of an organism do not occur at the physicochemical level but at a higher level of integration. And finally, it is the genetic program which controls the development and activities of the organic integrons that emerge at each successively higher level of integration.

The Distinguishing Characteristics of Life

Today, whether one consults working biologists or philosophers of science, there seems to be a consensus on the nature of living organisms. At the molecular level, all—and at the cellular level, most—of their functions obey the laws of physics and chemistry. There is no residue that would require autonomous vitalist principles. Yet, organisms are fundamentally different from inert matter. They are hierarchically ordered systems with many emergent properties never found in inanimate matter; and, most importantly, their activities are governed by genetic programs containing historically acquired information, again something absent in inanimate nature.

As a result, living organisms represent a remarkable form of dualism. This is not a dualism of body and soul, or body and mind, that is, a dualism partly physical and partly metaphysical. The dualism of modern biology is consistently physicochemical, and it arises from the fact that organisms possess both a genotype and a phenotype. The genotype, consisting of nucleic acids, requires for its understanding evolutionary explanations. The phenotype, constructed on the basis of the information provided by the genotype, and consisting of proteins, lipids, and other macromolecules, requires functional (proximate) explanations for its understanding. Such duality is unknown in the inanimate world. Explanations of the genotype and of the phenotype require different kinds of theories.

We may tabulate some of the phenomena that are specific to living beings:

Evolved programs. Organisms are the product of 3.8 billion years of evolution. All their characteristics reflect this history. Development, behavior, and all other activities of living organisms are in part controlled by genetic (and somatic) programs that are the result of the genetic information accumulated throughout the history of life. Historically there has been an unbroken stream from the origin of life and the simplest prokaryotes up to gigantic trees, elephants, whales, and humans.

Chemical properties. Although ultimately living organisms consist of the same atoms as inanimate matter, the kinds of molecules responsible for the development and function of living organisms—nucleic acids, peptides, enzymes, hormones, the components of membranes—are macromolecules not found in inanimate nature. Organic chemistry and biochemistry have shown that all substances found in living organisms can be broken down into simpler inorganic molecules and can, at least in principle, be synthesized in the laboratory.

Regulatory mechanisms. Living systems are characterized by all sorts of control and regulatory mechanisms, including multiple feedback mechanisms, that maintain the steady state of the system, mechanisms of a sort never found in inanimate nature.

Organization. Living organisms are complex, ordered systems. This explains their capacity for regulation and for control of the interaction of the genotype, as well as their developmental and evolutionary constraints.

Teleonomic systems. Living organisms are adapted systems, the result of countless previous generations having been subjected to natural selection. These systems are programmed for teleonomic (goal-directed) activities from embryonic development to the physiological and behavioral activities of the adults.

Limited order of magnitude. The size of living organisms occupies a limited range in the middle world, from the smallest viruses to the largest whales and trees. The basic units of biological organization, cells and cellular components, are very small, which gives organisms great developmental and evolutionary flexibility.

Life cycle. Organisms, at least sexually reproducing ones, go through a definite life cycle beginning with a zygote (fertilized egg) and passing through various embryonic or larval stages until adulthood is reached. The complexities of the life cycle vary from species to species, including in some species an alternation of sexual and asexual generations.

Open systems. Living organisms continuously obtain energy and materials from the external environment and eliminate the end-products of metabolism. Being open systems, they are not subject to the limitations of the second law of thermodynamics.

These properties of living organisms give them a number of capacities not present in inanimate systems:

A capacity for evolution

A capacity for self-replication

A capacity for growth and differentiation via a genetic program

A capacity for metabolism (the binding and releasing of energy)

A capacity for self-regulation, to keep the complex system in steady state (homeostasis, feedback)

A capacity (through perception and sense organs) for response to stimuli from the environment

A capacity for change at two levels, that of the phenotype and that of the genotype.

All these characteristics of living organisms distinguish them categorically from inanimate systems. The gradual recognition of this uniqueness and separateness of the living world has resulted in the branch of science called biology, and has led to a recognition of the autonomy of this science, as we will see in Chapter 2.

CHAPTER TWO

What Is Science?

Biology encompasses all of the disciplines devoted to the study of living organisms. Sometimes these disciplines are referred to as the life sciences—a useful term that distinguishes biology from the physical sciences, whose focus is on the inanimate world. The social sciences, political science, military science, and many others comprise yet other systematized bodies of knowledge, and in addition to these academic specialties, we frequently encounter Marxist science, Western science, feminist science, and such putative sciences as Christian science and creationist science. Why do all of these various disciplines call themselves "science"? What are the characteristics of a true science that distinguish it from other systems of thought? Does biology have these features?

It should be easy to answer these basic questions, one would think. Doesn't everybody know what science is? That this is not the case is evident when one studies not just the offerings of the popular press but also the enormous professional literature dealing with this question.[1] T. H. Huxley, a friend of Charles Darwin and a popularizer of Darwin's theories, defined science as "nothing but trained and organized common sense." Alas, this is not true. Common sense is frequently corrected by science. For instance, common sense tells us that the earth is flat and that the sun circles the earth. In every branch of science there have been commonsense opinions that have subsequently been proven wrong. One might go so far as to say that scientific activity consists of either confirming or refuting common sense.

A number of factors account for the difficulties philosophers have encountered in agreeing upon a definition of science. One of them is that science is both an activity (that which scientists do) and a body of knowledge (that which scientists know). Most philosophers today, in their definition of science, emphasize the ongoing activity of scientists: exploration, explanation, and testing. But other philosophers tend to define science as a growing body of knowledge, "the organization and classification of knowledge on the basis of explanatory principles."[2]

Emphasis on the collection of data and the accumulation of knowledge is a residue of the early days of the Scientific Revolution, when induction was the preferred method of science. There was a widespread misconception among inductionists that a pile of facts would not only permit generalizations but almost automatically produce new theories, as if by spontaneous combustion. Actually, philosophers today generally agree that facts alone do not explain, and they even argue a great deal over the question whether pure facts exist at all. "Are not all observations theory-laden?" they have asked. Even this is not a new concern. As far back as 1861, Charles Darwin wrote, "How odd it is that anyone should not see that all observation must be for or against some view if it is to be of any service."

To be sure, most authors who use the word "knowledge" mean it to include not just facts but also an interpretation of the facts; it is less confusing, however, to use the word "understanding" for this meaning. Hence the definition, "The aim of science is to advance our understanding of nature." Some philosophers would add "by solving scientific problems."[3] Some have gone further and have said, "The aims of science are to understand, predict, and control." Yet there are many branches of science in which prediction plays a very subordinate role, and in many of the nonapplied sciences the question of control never comes up.

Another reason for the difficulties philosophers have had in agreeing on a definition of science is that the endeavors which we call science have changed continually over the centuries. For example, natural theology—the study of nature for the purpose of understanding God's intentions—was considered a legitimate branch of science until about 150 years ago. As a result, in 1859 some of Darwin's critics chided him for including in his account of the origin of species such an "unscientific" factor as chance, while ignoring what they clearly saw as the hand of God in the design of all creatures great and small. Yet in the twentieth century we have witnessed a complete reversal in scientists' view of random phenomena: in both the life sciences and the physical sciences there has been a change from a strictly deterministic notion of how the natural world works to a conception that is largely probabilistic.

To take another example of how science is gradually changing, the strong empiricism of the Scientific Revolution led to a heavy emphasis on the discovery of new facts, while curiously little reference was made to the important role that the development of new concepts plays in the advancement of science. Today, concepts such as competition, common descent, territory, and altruism are as significant in biology as laws and discoveries are in the physical sciences, and yet their importance was strangely ignored until quite recently. This neglect is reflected, for example, in the provisions established for Nobel Prizes. Even if there were a Nobel Prize in biology (which there is not), Darwin could not have been awarded a prize for the development of the concept of natural selection—surely the greatest scientific achievement of the nineteenth century—because it was not a discovery. This attitude which favors discoveries over concepts continues into the present day, but to a lesser extent than in Darwin's time.

No one knows what other changes in our image of science the future may bring. The best one can do under the circumstances is to try to present an outline of the kind of science that prevails in our time, at the end of the twentieth century.

The Origins of Modern Science

Modern science began with the Scientific Revolution, that remarkable achievement of the human intellect characterized by the names Copernicus, Galileo, Kepler, Newton, Descartes, and Leibniz. At that time many of the basic principles of the scientific method were developed, which still largely characterize science today. What one considers science is, of course, a matter of opinion. In some respects Aristotle's biology was also science, but it lacked the methodological rigor and comprehensiveness of the science of biology as it developed from 1830 to the 1860s.

The scientific disciplines that gave rise to the prevailing concept of science during the Scientific Revolution were mathematics, mechanics, and astronomy. How large a contribution scholastic logic made to the original framework of this physicalist science has not yet been fully determined; it certainly played a major role in Descartes's thinking. The ideals of this new, rational science were objectivity, empiricism, inductivism, and an endeavor to eliminate all remnants of metaphysics—that is, magical or superstitious explanations of phenomena that were not grounded in the physical world.

Virtually all architects of the Scientific Revolution remained devout Christians, however; and, not surprisingly, the kind of science they created was very much a branch of the Christian faith. In this view, the world was created by God and thus it could not be chaotic. It was governed by His laws, which, because they were God's laws, were universal. An explanation of a phenomenon or process was considered to be sound if it was consistent with one of these laws. With the workings of the cosmos thus ultimately clear-cut and absolute, it should be possible eventually to prove and predict everything. The task of God's science, then, was to find these universal laws, to find the ultimate truth of everything as embodied in these laws, and to test their truth by way of predictions and experiments.

As far as mechanics was concerned, matters conformed rather well to this ideal. Planets orbited the sun and balls rolled down inclined planes in a predictable manner. Perhaps it was not an accident of history that mechanics, being the simplest of all sciences, was the first to develop a set of coherent laws and methods. But as the other branches of physics developed, exceptions to the universality and determinacy of mechanics were found again and again, requiring various modifications. Indeed, in everyday life the laws of mechanics are often so completely thwarted by random (stochastic) processes that determinacy appears to be totally absent. For instance, so much turbulence usually accompanies the movement of air masses and water masses that the laws of mechanics do not permit long-term predictions in either meteorology or oceanography.

The mechanists' recipe for the natural world worked even less well for the biological sciences. There was no room in the scientific method of the mechanists for the reconstruction of historical sequences, as occurred in the evolution of life, nor for the pluralism of answers and causations that make prediction of the future in the biological sciences impossible. When evolutionary biology was examined for its "scientificness" according to the criteria of mechanics, it flunked the test.

This was particularly true when it came to the favorite investigative method of mechanics: the experiment. The experiment was so valuable in this field that eventually it came to be treated as if it were the *only* valid scientific method. Any other method was considered inferior science. But since it was not in good taste to call one's colleagues bad scientists, these other nonexperimental sciences came to be called descriptive sciences. This term was for centuries pejoratively attached to the life sciences.

Actually, our basic knowledge in *all* sciences is based upon description. The younger a science is, the more descriptive it has to be to lay a factual foundation. Even today, most publications in molecular biology are essentially descriptive. What is really meant by "descriptive" is "observational," for all description is based on observation, whether by the naked eye or other sense organs, by simple microscopes or telescopes, or by means of highly sophisticated instrumentation. Even during the Scientific Revolution, observation (rather than experimentation) played a decisive role in the advance of science. The cosmological generalizations of Copernicus, Kepler, and for the most part Newton were based on observation rather than on laboratory experiments. Today, the underlying theories in fields such as astronomy, astrophysics, cosmology, planetary science, and geology change frequently as a result of new observations that have little if anything to do with experimentation.

One might put it another way and say that the findings described by Galileo and his followers came from the experiments of nature they were able to observe. The eclipses and occlusions of planets and stars are natural experiments, as are earthquakes, volcanic eruptions, meteor craters, magnetic shifts, and erosion events. In evolutionary biology, the joining of North and South America in the Pliocene through the Isthmus of Panama, which resulted in a massive faunal interchange of the two continents, is one such experiment; the colonization of volcanic islands and archipelagos such as Krakatau, the Galapagos, and the Hawaiian Islands, not to mention the defaunation and subsequent recolonization of much of the northern hemisphere owing to the Pleistocene glaciations, are other natural experiments. Much progress in the observational sciences is due to the genius of those who have discovered, critically evaluated, and compared such natural experiments in fields where a laboratory experiment is highly impractical, if not impossible.

A revolution in thought though the Scientific Revolution was—by abandoning superstition, magic, and the dogmas of medieval theologians—it nevertheless did not include a revolt against allegiance to the Christian religion, and this ideological bias had adverse consequences for biology. The answer to the most basic problems in the study of living organisms depends on whether or not one invokes the hand of God. This is particularly true for all questions of origin (the subject matter of interest to creationists) and design (the subject matter of interest to natural theologians). The acceptance of a universe containing nothing but God, human souls, matter, and motion worked fine for the physical sciences of the day, but it worked against the advance of biology.[4]

As a result, biology was basically dormant until the nineteenth and twentieth centuries. Although a considerable amount of factual knowledge in natural history, anatomy, and physiology was accumulated during the seventeenth and eighteenth centuries, the world of life at that time was considered to belong to the realm of medicine; this was true for anatomy and physiology, indeed, even for botany, which largely consisted of the identification of medicinally important plants. To be sure, there was also some natural history, but either it was practiced as a hobby or it was pursued in the service of natural theology. In retrospect, it is evident that some of this early natural history was very good science; but, not being recognized as such at that time, it did not contribute to the philosophy of science.

Finally, the acceptance of mechanics as the exemplar of science led to the belief that organisms are in no way different from inert matter. From this followed logically the conclusion that the goal of science was to reduce all of biology to the laws of chemistry and physics. In due time developments in biology made this position untenable (see Chapter 1). The eventual overthrow of mechanicism and its nemesis, vitalism, and the acceptance in the twentieth century of the paradigm of organicism have had a profound impact on the position of biology among the sciences—an impact not yet fully appreciated by many philosophers of science.

Is Biology an Autonomous Science?

After the middle of the twentieth century, one could discern three very different views on the position of biology in the sciences. According to one extreme, biology is to be excluded from science altogether because it lacks the universality, the law-structuredness, and strictly quantitative nature of a "true science" (meaning physics). According to the other extreme, biology not only has all the necessary attributes of a genuine science but differs from physics in important respects so that it is to be ranked as an autonomous science, equivalent to physics. Between these two extremes is the view that biology should be accorded the status of a "provincial" science, because it lacks universality and because its findings can ultimately be reduced to the laws of physics and chemistry.

The question "Is biology an autonomous science?" can be rephrased in two parts: "Is biology, like physics and chemistry, a science?" and "Is biology a science exactly like physics and chemistry?" To answer the first question, we might consult John Moore's eight criteria for determining whether a certain activity qualifies as science. According to Moore (1993): (1) A science must be based on data collected in the field or laboratory by observation or experiment, without invoking supernatural factors. (2) Data must be collected to answer questions, and observations must be made to strengthen or refute conjectures. (3) Objective methods must be employed in order to minimize any possible bias. (4) Hypotheses must be consistent with the observations and compatible with the general conceptual framework. (5) All hypotheses must be tested, and, if possible, competing hypotheses must be developed, and their degree of validity (problem-solving capacity) must be compared. (6) Generalizations must be universally valid within the domain of the particular science. Unique events must be explicable without invoking supernatural factors. (7) In order to eliminate the possibility of error, a fact or discovery must be fully accepted only if (repeatedly) confirmed by other investigators. (8) Science is characterized by the steady improvement of scientific theories, by the replacement of faulty or incomplete theories, and by the solution of previously puzzling problems.

Judging by these criteria, most people would conclude that biology should be considered, like physics and chemistry, a legitimate science. But is biology a provincial science, and therefore not on a par with the physical sciences? When the term "provincial science" was first introduced, it was used as an antonym to "universal," meaning that biology dealt with specific and localized objects about which one could not propose universal laws. The laws of physics, it was said, have no limitations of time or space; they are as valid in the Andromeda galaxy as on earth. Biology, by contrast, is provincial because all life that we know of has existed only on the earth, and only for 3.8 billion of the 10 billion or more years since the Big Bang.

This argument was convincingly refuted by Ronald Munson (1975), who showed that none of the fundamental laws, theories, or principles of biology are either implicitly or explicitly restricted in their scope or range of application to a certain region of space or time. There is a great deal of uniqueness in the world of life, but one can make all sorts of generalizations about unique phenomena. Each ocean current is also unique, but we can establish laws and theories about ocean currents. As for the argument that the restriction of known life to the earth deprives biological principles of all universality, here we must ask, "What is 'universal'?" Since inanimate matter is known to exist outside the earth, any science dealing with inanimate matter must be applicable extraterrestrially in order to be universal. Life, so far, has been demonstrated for the earth only; yet its laws and principles (like those of inanimate matter) are universal because they are valid on the earth, the known domain of its existence. I can see no reason for withholding the designation "universal" from a principle that is true for the entire domain for which it is applicable.

More often, when biology is described as a "provincial" science, what is meant is that it is a subset of physics and chemistry, and that ultimately the findings of biology can be reduced to chemical and physical theories. By contrast, an advocate of the autonomy of biology might argue in the following way: Many attributes of living organisms that interest biologists cannot be reduced to physicochemical laws, and, moreover, many aspects of the physical world studied by physicists are not relevant to the study of life (or to any other science outside of physics). In this sense physics is as provincial a science as biology. There is no reason to consider physics as an exemplar merely because it was the first well-organized science. That historical fact does not make it any more universal than its younger sibling, biology. A unity of science cannot be achieved until it is accepted that science contains a number of separate provinces, one of which is physics, another of which is biology. It would be futile to try to "reduce" biology, one provincial science, to physics, another provincial science, or vice versa.[5]

Many, if not most, of the promoters of the unity of science movement in the late nineteenth and early twentieth centuries were philosophers rather than scientists and had little awareness of the heterogeneity of the sciences. This applies to the physical sciences—which include elementary particle physics, solid state physics, quantum mechanics, classical mechanics, relativity theory, electromagnetism, not to mention geophysics, astrophysics, oceanography, geology, and others—and increases exponentially when we think of the many life sciences. The impossibility of reducing all these domains to a single common denominator has been demonstrated again and again during the past 70 years.

So to reiterate: Yes, biology is, like physics and chemistry, a science. But biology is not a science like physics and chemistry; it is rather an autonomous science on a par with the equally autonomous physical sciences. Nevertheless, one would not be able to speak of science in the singular if not all sciences, in spite of their unique features and a certain amount of autonomy, did not share common features. One of the tasks of the philosopher of biology is to establish what the common features are which biology shares with the other sciences, not only in methodology but also in principles and concepts. And these common features would define a unified science.

The Concerns of Science

It has been said that the scientist searches for truth, but many people who are not scientists claim the same. The world and all that is in it are the sphere of interest not only of scientists but also of theologians, philosophers, poets, and politicians. How can one make a demarcation between their concerns and those of the scientist?

HOW SCIENCE DIFFERS FROM THEOLOGY

The demarcation between science and theology is perhaps easiest, because scientists do not invoke the supernatural to explain how the natural world works, and they do not rely on divine revelation to understand it. When early humans tried to give explanations for natural phenomena, particularly for disasters, invariably they invoked supernatural beings and forces, and even today divine revelation is as legitimate a source of truth for many pious Christians as is science. Virtually all scientists known to me personally have religion in the best sense of this word, but scientists do not invoke supernatural causations or divine revelation.

Another feature of science that distinguishes it from theology is its openness. Religions are characterized by their relative inviolability; in revealed religions, a difference in the interpretation of even a single word in the revealed founding document may lead to the origin of a new religion. This contrasts dramatically with the situation in any active field of science, where one finds different versions of almost any theory. New conjectures are made continuously, earlier ones are refuted, and at all times considerable intellectual diversity exists. Indeed, it is by a Darwinian process of variation and selection in the formation and testing of hypotheses that science advances (see Chapter 5).

Despite the openness of science to new facts and hypotheses, it must be said that virtually all scientists—somewhat like theologians—bring a set of what we might call "first principles" with them to the study of the natural world. One of these axiomatic assumptions is that there *is* a real world, independent of human perceptions. This might be called the principle of objectivity (as opposed to subjectivity) or commonsense realism (see Chapter 3). This principle does not mean that individual scientists are always "objective" or even that objectivity among human beings is possible in any absolute sense. What it does mean is that an objective world exists outside of the influence of subjective human perception. Most scientists—though not all—believe in this axiom.

Second, scientists assume that this world is not chaotic but is structured in some way, and that most, if not all, aspects of this structure will yield to the tools of scientific investigation. A primary tool used in all scientific activity is testing. Every new fact and every new explanation must be tested again and again, preferably by different investigators using different methods (see Chapters 3 and 4). Every confirmation strengthens the probability of the "truth" of a fact or explanation, and every falsification or refutation strengthens the probability that an opposing theory is correct. One of the most characteristic features of science is this openness to challenge. The willingness to abandon a currently accepted belief when a new, better one is proposed is an important demarcation between science and religious dogma.

The method used to test for "truth" in science will vary depending on whether one is testing a fact or an explanation. The existence of a continent of Atlantis between Europe and America became doubtful when no such continent was discovered during the first few Atlantic crossings in the period of discoveries during the late fifteenth and early sixteenth centuries. After complete oceanographic surveys of the Atlantic Ocean were made and, even more convincingly, after photographs from satellites were taken in this century, the new evidence conclusively proved that no such continent exists. Often, in science, the absolute truth of a fact can be established. The absolute truth of an explanation or theory is much harder, and usually takes much longer, to gain acceptance. The "theory" of evolution through natural selection was not fully accepted as valid by scientists for over 100 years; and even today, in some religious sects, there are people who do not believe it.

Third, most scientists assume that there is historical and causal continuity among all phenomena in the material universe, and they include within the domain of legitimate scientific study everything known to exist or to happen in this universe. But they do not go beyond the material world. Theologians may also be interested in the physical world, but in addition they usually believe in a metaphysical or supernatural realm inhabited by souls, spirits, angels, or gods, and this heaven or nirvana is often believed to be the future resting place of all believers after death. Such supernatural constructions are beyond the scope of science.

HOW SCIENCE DIFFERS FROM PHILOSOPHY

The demarcation between science and philosophy is more difficult to determine than that between science and theology, and this led to tension between scientists and philosophers throughout most of the nineteenth century. Philosophy and science were a single endeavor at the time of the Greeks. The beginning of a separation of the two took place in the Scientific Revolution; but right up to Immanuel Kant, William Whewell, and William Herschel, many people who contributed to the advance of science were also philosophers. Later authors, like Ernst Mach or Hans Driesch, started out as scientists and then switched to philosophy.

Is there, perhaps, no demarcation at all between science and philosophy? The search for and discovery of facts is surely the business of science; but elsewhere there is a considerable area of overlap. Theorizing, generalizing, and establishing a conceptual framework for their field is considered by most scientists to be part of their job; indeed, it is this that makes the real scientist. Yet many philosophers of science have felt that theorizing and concept formation are the domain of philosophy. For better or for worse, in recent decades most of this endeavor has now been taken over by scientists, and some basic concepts developed by biologists have subsequently been taken up by philosophers and are now also concepts of philosophy.

To replace their former chief concern, philosophers of science have specialized in elucidating the principles whereby theories or concepts are formed. They search for the rules that specify the operations by which scientists answer the "What?" "How?" and "Why?" questions they encounter. The major domain of philosophy relating to science is now the testing of "the logic of justification" and the methodology of explanation (see Chapter 3). At its worst, this type of philosophy tends to degenerate into logic-chopping and semantic quibbling. At its best, it has forced scientists into responsibility and precision.

Although philosophers of science often state that their methodological rules are merely descriptive and not prescriptive, many of them seem to consider it their task to determine what scientists *should* be doing. Scientists usually pay no attention to this normative advice but rather choose that approach which (they hope) will lead most quickly to results; these approaches may differ from case to case.

Perhaps the greatest failing of the philosophy of science, until only a few years ago, was that it took physics as the exemplar of science. As a result, the so-called philosophy of science was nothing but a philosophy of the physical sciences. This has changed under the influence of the younger philosophers, many of whom specialize in the philosophy of biology. The intimate connection that exists today between philosophy and the life sciences is evident from the many articles published in the journal *Biology and Philosophy.* Through the efforts of these young philosophers, the concepts and methods used in the biological sciences have now become important components of the philosophy of science.

This is a most desirable development for both philosophy and biology. It should be the aim of every scientist to eventually generalize his views of nature so that they make a contribution to the philosophy of science. As long as the philosophy of science was restricted to the laws and methods of physics, it was not possible for biologists to make such a contribution. Fortunately, this is no longer the case.

The incorporation of biology has modified many of the tenets of the philosophy of science. As we will see in Chapters 3 and 4, the rejection of strict determinism and of reliance on universal laws, the acceptance of merely probabilistic prediction and of historical narratives, the acknowledgment of the important role of concepts in theory formation, the recognition of the population concept and of the role of unique individuals, and many other aspects of biological thought have affected the philosophy of science fundamentally. With probabilism now dominant, all aspects of logical analysis that are based on typological assumptions have become highly vulnerable. The complete certainty which, following Descartes, had been the ideal of the philosophers of science seems less and less important as a goal.

HOW SCIENCE DIFFERS FROM THE HUMANITIES

As far as the demarcation between science and the humanities is concerned, the tendency of writers in the past to ignore the heterogeneity of both fields has led to many misconceptions. There is more difference between physics and evolutionary biology—both of which are branches of science—than between evolutionary biology (one of the sciences) and history (one of the humanities). Literary criticism has virtually nothing in common with most of the other disciplines of the humanities and even less with science.

When C. P. Snow wrote his *Two Cultures* in 1959, what he actually described was the gap between physics and the humanities. Like others of that era, he naively assumed that physics could stand for science as a whole. The gap between physics and the humanities, as he rightly pointed out, is indeed virtually unbridgeable. There is simply no pathway from physics to ethics, culture, mind, free will, and other humanistic concerns. The absence in physics of these important topics contributed to the alienation of scientists and humanists that Snow decried. Yet, all these concerns have substantial relationships with the life sciences.

Similarly, when E. M. Carr (1961), a humanist, contrasted history with "the sciences," he found five respects in which they differ: (1) History, he said, deals exclusively with the unique, science with the general. (2) History teaches no lessons. (3) History, unlike science, is unable to predict. (4) History is necessarily subjective, while science is objective. (5) And history, unlike science, touches upon issues of religion and morality. What Carr failed to see was that these differences are valid only for the physical sciences and for much of functional biology. However, statements 1, 3, and 5 apply as well to evolutionary biology as to history, and, as Carr admits, some of these claims (statement 2, for instance) are not strictly true even for history. In other words, the sharp break between the "sciences" and the "nonsciences" does not exist, once biology is admitted into the realm of science.[6]

Quite often the estrangement between science and the humanities is assigned to the failure of scientists to appreciate the "human element" as they go about their research. Yet not all of the blame should be shouldered by scientists. A rudimentary knowledge of certain findings of science, particularly of evolutionary biology, behavioral science, human development, and physical anthropology, is indispensable for most work in the humanities. Yet, all too many humanists have failed to acquire such a knowledge and display an embarrassing ignorance of these subjects in their writings. Many excuse their poor understanding of science with the statement, "I have no ability in mathematics." Actually, there is little mathematics in those parts of biology with which the humanists should most familiarize themselves. For instance, there is not a single mathematical formula in Darwin's *Origin of Species* or in my *Growth of Biological Thought* (1982). An understanding of human biology should be a necessary and inseparable component of studies in the humanities. Psychology, formerly classified with the humanities, is now considered a biological science. Yet, how can one write anything in the humanities, whether in history or literature, without having a considerable understanding of human behavior?

Snow correctly emphasized this point. There is a deplorable ignorance among most people of even the simplest facts of science. For example, writer after writer still states that he cannot believe that the eye is the result of a series of accidents. What this statement reveals is that the writer has no understanding of the workings of natural selection, which is an *anti*chance—rather than an accidental—process. Evolutionary change occurs because certain characteristics of individuals are better suited to the current environmental circumstances of a species than are others, and these more adaptive features become concentrated in later generations through differential rates of survival and reproduction—in other words, through selection. Chance certainly plays a part in evolution, as Darwin knew very well, but natural selection—the primary mechanism of evolutionary change—is not an accidental process.

An ignorance of the findings of biology is particularly damaging whenever humanists are forced to confront such political problems as global overpopulation, the spread of infectious diseases, the depletion of nonrenewable resources, deleterious climatic changes, increased agricultural requirements worldwide, the destruction of natural habitats, the proliferation of criminal behavior, or the failures of our educational system. None of these problems can be satisfactorily addressed without taking into account the findings of science, particularly biology, and yet too often politicians proceed in ignorance.

The Objectives of Scientific Research

It is often asked why do we do science? Or, what is science good for? Two rather different answers to this question have been given. The insatiable curiosity of human beings, and the desire for a better understanding of the world they live in, is the primary reason for an interest in science by most scientists. It is based on the conviction that none of the philosophical or purely ideological theories of the world can compete in the long run with the understanding of the world produced by science.

To make a contribution to this better understanding of the world is a source of great satisfaction to a scientist; indeed it is an occasion for exhilaration. The emphasis is often on discovery, where luck sometimes plays a role, but the joy is perhaps even greater when one succeeds in the difficult intellectual achievement of developing a new concept, a concept that can integrate a mass of previously disparate facts, or one that is more successful as the basis of scientific theories. Offsetting the joy of research, of course, is the incessant need for dull data-collecting, the disappointment (if not embarrassment) of invalid theories, the recalcitrance of certain research subjects, and a multitude of other frustrations.[7]

An entirely different objective is to use science as a means to control the world, its forces and resources. This second objective is held particularly by applied scientists (including those in medicine, public health, and agriculture), engineers, politicians, and the average citizen. But what some politicians and voters forget is that when it comes to the ills of pollution, urbanization, famine, or the population explosion, it is not sufficient to fight the symptoms. One does not cure malaria with aspirin, and one cannot fight social and economic ills without going into the causes. Our way of dealing with racial discrimination, crime, drug addiction, homelessness, and similar problems, and the success we will have in eliminating them, will depend to a considerable extent on our understanding of their biological roots.

These two objectives of science—satisfying curiosity and making improvements in the world—are not entirely different domains, because even applied science, particularly all science on which public policy is based, relies on basic science. In most cases scientists are largely motivated by the simple desire for a better understanding of puzzling phenomena in our world.

In both basic and applied science, any discussion of the objectives of scientific research always entails questions of values. To what extent can our society afford certain big science projects, like the superconducting supercollider or the space station, considering the narrowness of the results we can expect to obtain? To what extent should one consider certain experiments, particularly with mammals (dogs, monkeys, apes) as unethical? Is there a danger that work with human embryonic materials might lead to unethical practices? What experiments in human psychology or clinical medicine might be harmful to the experimental subjects?

As long as the physical sciences were dominant, science was usually considered to be value-free. During the student rebellion of the 1960s, some groups who resented this arrogance promoted the slogan "Down with value-free science." Since the rise of biology, and particularly of genetics and evolutionary biology, it has become clear that scientific findings and theories have an impact on values, though to what extent science can generate values is unclear (see Chapter 12). Some of Darwin's opponents, such as Adam Sedgwick, accused Darwinism of destroying moral values. Even today, creationists fight evolutionary biology because they are convinced that it undermines the values of Christian theology. The eugenics movement in this century clearly derived its values from the science of human genetics. And the reason why sociobiology was attacked so viciously in the 1970s was that it seemed to promote certain political values incompatible with those of its opponents. Almost all major religious and political ideologies uphold values that are claimed to derive from science, and almost all ideologies uphold other values that are incompatible with certain findings of science.

Paul Feyerabend (1970) has ventured to suggest (as have other contemporary writers) that a world without science "would be more pleasant than the world we live in today." I am not sure that this is true. There would be less pollution and pollution-caused cancer, less crowding, and fewer of the adverse by-products of mass society. But it would also be a world with high infant mortality, a life span of only 35–40 years, no way of escaping summer heat and protecting oneself against severe winter cold. It is all too easy to forget the vast benefits of science (including agricultural and medical science) when one is complaining about its deleterious side effects. Most of these so-called evils of science and technology could be eliminated; scientists know what should be done, but their knowledge must be translated into legislation and its enforcement, and this has so far been resisted by the politicians and much of the voting public.

My own view of the contributions of science is more in line with that of Karl Popper, who had this to say: "Next to music and art, science is the greatest, most beautiful and most enlightening achievement of the human spirit. I abhor the at present so noisy intellectual fashion that tries to denigrate science, and I admire beyond anything the marvelous results achieved in our time by the work of biologists and biochemists and made available through medicine to sufferers all over our beautiful earth."

SCIENCE AND THE SCIENTIST

One frequently hears that science can do this, or science cannot do that, but of course it is scientists who either can or cannot do something. A scientist at his or her best is dedicated, highly motivated, scrupulously honest, generous, and cooperative. Scientists are only human, however, and do not always live up to these professional ideals. Political, theological, or financial considerations that arise from outside of science should not, but often do, affect scientific judgment.

Scientists have their own specific traditions and values, which they learn from a mentor, older colleague, or other role model. This includes not only the avoidance of dishonesty or fraud but also giving appropriate credit to competitors if they have priority in making a discovery. A good scientist will tenaciously defend his own priority claims, but at the same time he is usually anxious to please the leaders in his field and will sometimes follow their authority even when he should be more critical.

Any cheating or manufacturing of data is discovered sooner or later and is the end of a career; for that reason alone, fraud is not a viable option in science. Inconsistency is perhaps a more widespread failing; there is probably no scientist who entirely escapes it. Charles Lyell, whose *Principles of Geology* influenced Darwin's thinking, preached uniformitarianism, but it struck even some of his contemporaries how nonuniformitarian was his own theory of the origin of new species. Darwin himself was also capable of inconsistency; he applied population thinking when explaining adaptation by natural selection, but he employed typological language in some of his discussions of speciation. Lamarck proclaimed loudly that he was a strict mechanist, endeavoring to explain everything in terms of mechanical causes and forces, and yet his discussion of inevitable perfection through evolutionary change strikes the modern reader as a subconscious adherence to a (nonmechanistic) perfecting principle. None of Darwin's adherents stressed natural selection more forcefully than A. R. Wallace, but when it came to applying it to man, Wallace "chickened out."

Some flaws in the findings and hypotheses of scientists are clearly induced by wishful thinking. When an early investigator found 48 chromosomes in the human species, this discovery was subsequently confirmed by numerous other investigators because that is the number they expected to find. The correct number (46) was not established until three different new techniques had been introduced.

Recognizing that error and inconsistency are widespread in science, Karl Popper in 1981 proposed a set of professional ethics for the scientist. The first principle is that there is no authority; scientific inferences go well beyond what any one person can master, including specialists. Second, all scientists at all times commit errors; they seem to be unavoidable. One should search for errors, analyze them when found, and learn from them; it is an unforgivable sin to conceal errors. Third, while such self-criticism is important, it must be supplemented by criticism by others, who can help discover and correct one's errors. In order to be able to learn from one's errors, one must acknowledge them when others call attention to them. And finally, one always must be aware of one's own errors when calling attention to those of others.

The major reward of a scientist is his prestige among his peers. This prestige depends on such factors as how many important discoveries he has made and what his contribution to the conceptual structure of his discipline has been. Why are priority and recognition by peers so important to most scientists? Why do a few scientists try to denigrate their peers (or competitors)? How is a scientist rewarded for achievements? What is the relationship of scientists to one another, and the relation of scientists to the rest of society? All such questions have been asked by researchers in the sociology of science, most importantly by Robert Merton, who virtually founded the discipline. As Merton has shown, much modern science is done by research groups, and alliances are often formed under the flag of certain dogmas.[8] But despite a certain degree of dissension in science, what impresses outsiders most is the remarkable consensus among scientists in the last half of the twentieth century.

This consensus is particularly well reflected in the internationality of science. English is rapidly becoming the lingua franca of science, and in certain countries, such as Scandinavia, Germany, and France, prominent scientific journals have adopted English names and publish primarily English-language articles. A scientist traveling to another country, even an American visiting Russia or Japan, feels quite at home when in the company of colleagues from those countries. Numerous articles are published these days in scientific journals in which the coauthors are from different countries. One hundred years ago scientific papers and books very often had a distinctly national flavor, but this is becoming rarer all the time.

All scientists who reach worthwhile goals tend to be ambitious and hard-working. There is no such thing as a 9-to-5 scientist. Many work 15 to 17 hours a day, at least during certain periods of their career. Yet most of them have broad interests, as is evident from their biographies; quite a number of scientists are amateur musicians, for instance. In other respects scientists are as variable a lot as any human group. Some are extroverts, others shy introverts. Some are exceedingly prolific, while others concentrate on the production of a few major books or papers. I do not think that there is a definite temperament or personality that one could identify as the typical scientist.

Traditionally one became a biologist either through a medical education or by growing up as a young naturalist. At present it is much more common for a youngster to become excited about the life sciences through the media, particularly nature films on television, visits to a museum (often the dinosaur hall), or an inspiring teacher. There are also thousands of young bird watchers, some of whom will become professional biologists (as I did). The most important ingredient is a fascination with the wonders of living creatures. And this stays with most biologists for their entire life. They never lose the excitement of scientific discovery, whether empirical or theoretical, nor the love of chasing after new ideas, new insights, new organisms. And so much in biology has a direct bearing on one's own circumstances and personal values. Being a biologist does not mean having a job; it means choosing a way of life.[5]

CHAPTER THREE

How Does Science Explain
the Natural World?

The earliest attempts to explain the natural world invoked the supernatural. From the most primitive animism to the great monotheistic religions, anything that was puzzling and seemingly inexplicable was attributed to the activities of spirits or gods. The ancient Greeks initiated a different approach. They attempted to explain the phenomena of the world through natural forces. Philosophy, which developed in the sixth century BC, became occupied increasingly with the task of explaining the world and attempting to determine what the ideal of "knowing" should be. The Greeks based their explanations on observation and thinking, though metaphysics always played a considerable role. From these early beginnings, the philosophy of science that we recognize today gradually developed.

The third kind of explanatory endeavor was science, which arose during the Scientific Revolution. Supernatural explanations, philosophy, and science are perhaps best considered not three consecutive stages but rather three complementary approaches to the problem of knowing. The history of human thought shows that these differing endeavors evolved from one another without sharp breaks. For example, many of the great philosophers, even Kant, included God in their explanatory schemes. Prior to Darwin, God was also accepted as an explanatory factor by most biologists. After the rise of science, philosophy continued to exist and prosper; what changed was its objective. As science gradually became emancipated from philosophy, philosophers began to stand back reflectively from the work of scientists and to focus on analysis of scientists' activities.

The ultimate aim of science is to advance our understanding of the world—on that point both scientists and philosophers of science agree. The scientist raises questions about that which is not known or not understood and attempts to answer them. The first answer is called a conjecture or hypothesis and serves as a tentative explanation. But what really is an explanation? When a puzzling phenomenon is encountered in the everyday world, most frequently it is "explained" in terms of what is known or what is rational. For example, an eclipse of the moon must be due to the earth's shadow falling on the moon, or the fauna and flora of the Galápagos Islands must have gotten there by overwater dispersal, because these volcanic islands obviously never had any connection with the South American continent. But merely having a rational explanation is not enough. One must also make sure that the answer is true, or at least as close to the truth as available knowledge permits. This goal of the scientist is precisely also the objective of the philosopher of science.

What has been controversial among philosophers from the age of the Greeks to modern times is how an explanation about the natural world should be constructed and tested. Scores of philosophers have endeavored to formulate principles by which our understanding of the world could be advanced (or, as it was often said, how truth could be found). Among those usually listed are Descartes, Leibniz, Locke, Hume, Kant, Herschel, Whewell, Mill, Jevons, Mach, Russell, and Popper. Curiously, the name of Darwin is rarely included in such a list, even though he was clearly one of the greatest philosophers of all times.[1] In fact, to a large extent the modern philosophy of biology was founded by Darwin.

Were these philosophers of science simply attempting to describe faithfully the methods of the scientist, as seen through the eyes of a philosopher, or was their endeavor to tell scientists how they should construct their explanations and tests so that their findings constitute truly "good" science?[2] If the latter is the case, I fear that so far it has had little effect. I do not know of a single biologist whose theorizing was much affected by the norms proposed by philosophers of science. Scientists usually go about their research without paying much attention to the fine points of methodology. The one exception is Karl Popper's insistence on falsification (see below), which was widely accepted by biologists in principle, though it rarely worked out in practice.

Why are philosophers of science still, today, so worried about the way that scientists construct and test their explanations? After all, science has had an almost unbroken series of successes ever since the Scientific Revolution. Of course, occasionally an erroneous theory is temporarily adopted, but it is soon refuted in the contest among competing theories. Cases of a refutation of a major scientific theory are remarkably rare. Overall, the reliability of the major claims of science is unquestionable. Giere (1988) suggests that the heritage of Cartesian skepticism during the Scientific Revolution is responsible for the continuing doubts of the philosophers.

The media, with their daily sensational announcements of major new discoveries and challenges to existing theories, tend to mislead the nonscientist into believing that science can yield no certainty or "truth" about anything. To the contrary, the basic theories of science, many of them as much as 50 or even 150 years old, are being confirmed again and again. Even in a field as controversial as evolutionary biology, the basic conceptual framework established by Darwin in 1859 has turned out to be remarkably robust. All attempts in the last 130 years to invalidate Darwinism (and there have been hundreds) have been unsuccessful, and the same is true for most other areas of biology.

Nevertheless it must be acknowledged that our sense organs are fallible and our reasoning even more so. It is therefore a legitimate task of philosophy to scrutinize the methods by which scientists obtain knowledge—indeed, to advise scientists as to the most reliable way to formulate and test theories. The branch of philosophy that deals with the problem of what we know and how we know it is called epistemology. It is the main preoccupation of the philosophy of science today.[3]

A Brief History of the Philosophy of Science

Not surprisingly, the rise of interest in epistemology coincided with, or was caused by, the Scientific Revolution. With astronomy and mechanics the most active sciences at that time, observation and mathematics were highly regarded, and Sir Francis Bacon (through induction) and Descartes (through geometry) became their apostles.

Through Bacon, induction became the established scientific method for two centuries. According to this philosophy, the scientist develops his theories by simply recording, measuring, and describing observations without having any prior hypotheses or preconceived expectations. When induction was fashionable in England in the early nineteenth century, Darwin proclaimed that he was a true follower of Bacon, while in reality what he adopted was a more or less hypothetico-deductive approach (see below).[4] Later Darwin made fun of induction, saying that if one did believe in this method, "one might as well go into a gravel pit, count the pebbles, and describe the colors."

Liebig (1863) was one of the first prominent scientists to repudiate Baconian induction, arguing convincingly that no scientist had ever, or could ever, follow the methods described in Bacon's *Novum Organum*. Induction by itself cannot generate new theories. Liebig's incisive critique helped to bring the reign of inductionism to an end,[5] and from then on it was considered derogatory to call someone an inductionist (or "stamp-collector"). Many of the critics of this empirical approach, however, overlooked the fact that the data underlying any scientific endeavor remained as indispensable as ever; what was to be criticized was not the collecting of facts per se but how these facts were used in theory formation. In some sciences (particularly in biology) which rely on the construction of historical narratives, the essential scientific method today is basically inductive.

Later in the nineteenth century, particularly under the influence of the work of Frege (1884) and of other logicians and mathematicians, logic became a dominant influence in the philosophy of mathematics and physics. This was particularly illuminating where mathematically formulated, universal laws played an important role, as in the physical sciences. It was less appropriate for biology, where pluralism, probabilism, and purely qualitative as well as historical phenomena abound, while strictly universal laws are virtually absent. As a consequence, a philosophy of science developed that was tailored for the situation in the physical sciences but was to a large extent inappropriate for biology.

VERIFICATION AND FALSIFICATION

In this century, the philosophy that long dominated Anglo-American science was logical empiricism, which grew out of the Vienna Circle of logical positivists (Reichenbach, Schlick, Carnap, Feigl) in the 1920s and 1930s. Logical empiricism was built on three foundations: (1) the work of a number of twentieth-century mathematicians and logicians; (2) the classical empiricism of David Hume as transmitted through Mill to Russell and Mach; and (3) the physical sciences, particularly the classical physical sciences as they were understood prior to relativity and quantum mechanics.

The approach to scientific confirmation endorsed by the logical positivists was the traditional hypothetico-deductive (H-D) method, and verification through repeated testing was considered to be the best criterion for the goodness of a theory. If testing confirms a theory, they would say that the theory had been verified. Verification greatly strengthens theories and sometimes leads to constructive modification. One must not assume, however, that verification "proves" unambiguously that a given theory is true. These methods have sometimes led to a verification of what ultimately turned out to be a wrong theory.[6]

Popper agreed with the logical positivists that a theory will be "regarded the more satisfactory the greater the severity of the independent tests it has survived," but he insisted that falsification was the only way to finally eliminate an invalid theory. If the theory fails a test, it has been falsified. Falsification is not a simple matter, however. It is not like proving that 2 + 2 is not 5. It is particularly ill-suited for the testing of probabilistic theories, which include most theories in biology. The occurrence of exceptions to a probabilistic theory does not necessarily constitute falsification. And in fields such as evolutionary biology, in which historical narratives must be constructed to explain certain observations, it is often very difficult, if not impossible, to decisively falsify an invalid theory. The categorical statement that a single falsification requires the abandonment of a theory might be true for theories based on the universal laws of the physical sciences, but is often not true for theories in evolutionary biology.[7]

NEW MODELS OF SCIENTIFIC EXPLANATION

The modern philosophy of science began in 1948 in a paper written by Carl Hempel and Paul Oppenheim and elaborated by Hempel in 1965. In these essays Hempel proposed a new model of scientific explanation, which he called the deductive-nomological (D-N) model. This schema had its heyday in the 1950s and 60s and was also known as the "received view."

The idea behind deductive-nomological explanation is this: A scientific explanation is a deductive argument in which a statement describing the to-be-explained event is deduced from one or more true universal laws in conjunction with statements of particular facts (correspondence rules). According to this view, a scientific theory is an "axiomatic deductive system," whose premises are based on a law.

The original D-N model was very typological and deterministic, and was soon modified to cope with probabilistic or statistical laws. Each year new papers or books were published suggesting ways and means to correct actual or seeming flaws in the received view. Some of these were proposed as genuinely new theories, although ultimately they were derived from the Hempel model.

One of these modifications became known as the semantic conception of theory structure.[8] For Beatty (1981, 1987), a proponent of this new model, a theory is the definition of a system, and applications of a theory are instantiations of theory. Such applications may or may not be spatio-temporally restricted. Theories are neither general nor permanent, and are therefore compatible with plural solutions and with evolutionary change. This last point is important in view of the fact that there are so few spatio-temporally unrestricted biological generalizations. The ability of the semantic view to faithfully represent evolutionary theorizing has induced Beatty, Thompson, Lloyd, and other philosophers to adopt the semantic view.[9]

Although this theory escapes several of the weaknesses of the received view, it faces two difficulties as far as the working biologist is concerned. The first is that when one asks for a definition of this approach, one gets exceedingly different versions from different semanticists. The second stumbling block is this: How can the semantic view be applied by the working biologist? What the philosopher offers is a description of theories that have been developed by the scientist. But such a description is not sufficiently normative to tell the biologist how to develop *new* theories. At least so it seems to me. When does a theory fail to measure up to the specifications of a semantic theory? Lack of an answer to this question perhaps is the reason why I do not feel that the semantic approach has achieved much acceptance in biology, in spite of its distinct advantages over the received view (which today is considered more or less obsolete). What has been increasingly appreciated is that the assessment of a theory is not a matter of simple logical rules and that rationality has to be construed in broader terms than either deductive or inductive logic offer.

Each of the various explanatory schemes of this century has had its vogue for ten years or more and then has been replaced by an amended version or an entirely new scheme.[10] The 1980s have been particularly active in the philosophy of science, but this activity has not led to any consensus among the philosophers on how best to construct and test a scientific explanation. In his recent survey, Salmon (1988) writes: "It seems to me that there are at least three powerful schools of thought at present—the pragmatists, the deductivists, and the mechanists—and that they are not likely to reach substantial agreement in the near future."

Discovery and Justification

Most scientists and philosophers of science seem basically to agree that science is a two-step process. The first step involves the *discovery* of new facts, irregularities, exceptions, or seeming contradictions in nature, and the formation of conjectures, hypotheses, or theories to explain them. The second step deals with *justification*—the procedures by which such theories are tested and validated.

For most philosophers, the pathway to a new theory begins with making a conjecture or hypothesis to solve a puzzle; this hypothesis is then subjected to rigorous testing. But the working scientist starts even earlier. During the discovery phase he engages in a great deal of simple observation and description of facts. When he encounters an unexplained irregularity or anomaly among the facts available to him, the discovery of this puzzle induces him to ask a question, and that question leads eventually to a conjecture or hypothesis.

Every scientist occasionally has "hunches" about the meaning or explanation of this or that observation. But it is only the successful testing of these hunches that moves scientific discovery to the stage of "truth." Justification—how one goes about testing conjectures, hypotheses, or theories—has become a preoccupation of philosophers of science, in large part because justification is amenable to logical analysis. Discovery only rarely follows "logically" from the preceding situation, and therefore most philosophers have traditionally not considered aspects of discovery to be their business. Rather, they usually ascribe discovery to chance, to psychological factors, to the Zeitgeist, or, worse, to prevailing socioeconomic conditions.

Popper (1968), for instance, asserted, "How it happens that a new idea occurs to a man . . . is irrelevant to the logical analysis of scientific knowledge. The latter is not concerned with questions of fact . . . but only with questions of justification or validity." Yet in the eyes of the working scientist, the method one uses to refute an erroneous hypothesis is usually of trivial interest, while the discovery of a new fact or the formulation of a new theory is frequently of the most fundamental significance.[11]

INTERNAL AND EXTERNAL FACTORS IN THEORY FORMATION

No scientist lives in a vacuum. He lives in an intellectual, spiritual, economic, and social, as well as scientific, environment. What impact do these influences have on the nature of the theories he develops? Intellectual historians tend to hold internal factors—that is, developments within science—as primarily responsible for new theories and concepts. Social historians, by contrast, search for external factors—that is, components of the socioeconomic milieu. On the whole the sociologists have been remarkably unsuccessful in their endeavor.[12] The fact that Charles Darwin and Alfred Russel Wallace, who came from

such totally different socioeconomic backgrounds, arrived independently at virtually the same theory of evolution illustrates the
irrelevance of external factors. Indeed, I know of no evidence whatsoever of the influence of a socioeconomic factor on the development
of a specific biological theory.[13] The reverse, however, is sometimes
true: scientific or pseudoscientific theories have frequently been used
by political activists to promote their particular agenda.[14]

Among the external factors, one must distinguish between socioeconomic factors and the Zeitgeist, or intellectual milieu. While the
latter seems to play only a small role in the proposal of new theories,
it seems to play a very large role in the resistance to intellectual shifts
that are in conflict with established beliefs. This was the reason why
Darwin's theory of natural selection encountered such massive resistance; in the conceptual world of Cuvier or Agassiz, it was impossible
to accommodate a theory of evolution.[15]

TESTING

How does a scientist go about determining whether his new hypothesis
is valid? By subjecting it to certain tests. The philosopher who wants
to determine the goodness of a theory does the same, but the testing
undertaken by scientists is sometimes quite different from that undertaken by philosophers, who tend to apply rules much more rigidly
than working scientists do.[16] Which set of rules will be applied differs,
however, depending on the school to which a given philosopher belongs.

For example, philosophers of science since the days of the logical
positivists have placed great stress on the capacity of theories to make
predictions. The better a theory is, the more correct the predictions
it permits. Prediction in this context means logical prediction: provided
that such and such a constellation of factors exists, one can expect
such and such an outcome to occur. This use of prediction in logic
is different from the everyday use of the word "prediction," which
means being able to foretell the future. Foretelling the future is *chronological prediction*. Many authors (including myself in the past) have
confused the two kinds of prediction. Science, often even the physical
sciences, is only rarely able to make chronological predictions. For

instance, nothing is as unpredictable as the future course of evolution. The dinosaurs were the most successful group of terrestrial vertebrates at the beginning of the Cretaceous; that they would be extinct by the end of that era, owing to the collision of an asteroid with the earth, was unpredictable.

The biologist, like the physicist, also applies the test of prediction and searches for exceptions, but he is less disturbed by the occasional failure of a prediction to come true because he knows that biological regularities rarely have the universality of physical laws. The usefulness of prediction in testing biological theories is highly variable. Some theories, particularly in functional biology, have high predictive value, whereas others are controlled by so complex a set of factors that consistent prediction is unachievable. Predictions in biology are at best probabilistic, owing to the great variability of most biological phenomena and owing to the occurrence of contingencies and the multiplicity of interacting factors that affect the course of events. For the biologist, it is not so important that his theory survive the test of prediction; it is more important that his theory is useful in solving problems.[17]

In the functional sciences, theories are best tested with the help of experiments. But in the sciences in which experiments are not possible and prediction is of limited value in testing a particular hypothesis—and this is usually the case in historical sciences—additional observations have to be made. For instance, the theory of common descent claims that the animals and plants of more recent geological periods are descendants of those from older geological periods. Giraffes and elephants, for instance, are descendants of early Tertiary taxa. It would discredit the theory of common descent if one found fossil elephants and giraffes in the early Cretaceous. Likewise, dinosaurs originated in the Mesozoic, and therefore it would contradict the theory of common descent if fossil dinosaurs were found in the Paleozoic.

Another way of testing a theory is to use an entirely different set of facts. For instance, if on the basis of morphological evidence I have constructed a phylogenetic tree of a certain group of organisms, I can use one of several types of molecular (biochemical) evidence to construct an independent phylogeny, and then test the degree of congruence of the two trees. Whenever there is a disagreement between the two trees, additional independent evidence must be used as a further check. In biogeography, theories about former land connections or about dispersal capacities of different taxa can be tested in various ways, and biogeographic theories can thus be refuted or strengthened. In order to prove that the dinosaurs truly became completely extinct at the end of the Cretaceous, additional early Tertiary deposits in remote areas of the world have to be examined. The nature of the observations and tests required are different from one problem to the next, though specialists are usually largely in agreement about what tests or observations should be considered valid in a given field.

The Practicing Biologist

None of the many philosophies of science proposed in this century—based as they were on laws and logic—has been well suited for theory development in evolutionary biology. This realization led Popper in 1974 to conclude, not that the scientific method prescribed is flawed, but "that Darwinism is not a testable scientific theory but a meta-physical research program." Other philosophers, also with a background in physics or mathematics, made similar statements. Popper recanted a few years later, and the philosophy of logical empiricism, after having been dominant for some 40 years, was abandoned owing to the critiques of Kuhn, Lakatos, Beatty, Laudan, Feyerabend, and other philosophers. In the long run, what logical empiricism managed to accomplish in the life sciences was to foster among many biologists a distrust in the philosophy of science.

Nevertheless, the average biologist, it seems to me, is not particularly worried over the state of affairs in the philosophy of science at any given time. When in the 1950s and 60s Popper was the great rage, every biologist I knew insisted that he was a Popperian, and then did whatever he wanted to do. Labels are sometimes politically convenient but they often mean nothing. (The situation reminds me of the story of the father who had two identical twin sons whom he could never tell apart. So he sent one to Harvard, the other to Yale. After four years the Harvard boy had become a typical refined Boston Brahmin, while the Yale boy had become a typical Yale bulldog—and the father still could not tell them apart.)

The working biologist does not ask whether he should follow the prescriptions of this or that school of philosophy. When one studies the history of various theories in science, one sympathizes with Feyerabend (1975), who claimed, "Anything goes." Indeed this attitude seems to be what guides the biologist in most of his theorizing. He does what François Jacob (1977)—with respect to natural selection—has called "tinkering." He uses whatever method will get him at the moment most conveniently to the solution of his problem.

FIVE STAGES OF EXPLANATION

In biology—where chance, pluralism, history, and uniqueness play such important roles (see Chapter 4)—a flexible system of theory construction and testing would seem more appropriate than rigid principles. Such a system might be captured in five words. (1) Scientists make *observations* on undisturbed nature, or during specifically directed experiments, some of which are unexplained by current theories or are in conflict with generally held views. (2) These observations lead the scientist to formulate *questions* of "How?" and "Why?" (3) To answer these questions, the investigator constructs a tentative *conjecture* or working hypothesis. (4) In order to determine if this conjecture is correct, he subjects it to rigorous *testing*, which will either strengthen the probability that it is valid, or weaken it; tests consist of making additional observations, preferably using different strategies or pathways as well as carefully designed experiments. (5) The *explanation* ultimately adopoted will be the conjecture that has been most successful during the testing procedure.

COMMONSENSE REALISM

Philosophers have endlessly speculated about whether there is a real world outside of us, as indicated by the stimuli received by our sense organs, and whether this world is exactly as we are told by our sense organs and by science. One extreme is represented by Bishop Berkeley's suggestion that the outside world is simply an outward projection from us.[18] Biologists known to me are commonsense realists. They accept it as a fact that a "real world" outside of us exists. We now have so many ways of testing our sense impressions by instrumentation, and the predictions based on such observations come true so invariably, that there would seem little benefit in challenging the pragmatic or commonsense realism on the basis of which biologists normally conduct their researches.

Common sense is not a fashionable tool among philosophers, who much prefer to rely on logic. To a nonlogician, by contrast, most syllogisms appear to be virtually identical equations. He is more comfortable with common sense. Also in the determination of the nature of causation, a commonsense approach is often the most comfortable and productive one. The rigorous approach of the logician might have been suitable for a deterministic, essentialistic world governed by universal laws, but it seems less appropriate in a probabilistic world ruled by contingencies and chance, a world in which one is forever asked to explain unique phenomena. White, pied, and brown ravens as well as black and black-necked swans (they all exist!) do not make a good case for a superiority of logic.

THE LANGUAGE OF SCIENCE

Each branch of science has its own terminology for the facts, processes, and concepts of its field. When a term refers to an object or individual—mitochondria, chromosomes, nucleus, gray wolf, Japanese beetle, dawn redwood—it usually poses no problem. But a large class of terms refer to more heterogeneous phenomena or processes; competition, evolution, species, adaptation, niche, hybridization, variety are some that are encountered in biology. When these terms are understood exactly the same way by all workers, they are helpful and indeed necessary.[19] However, as the history of science has shown, that is often not the case, and the result is misunderstanding and controversy.

Three kinds of problems with language are encountered by the working scientist. First, the meaning of a term may change as our knowledge of the subject grows. Such changes in meaning are not surprising, since scientific terms are usually borrowed from daily language and have all the vagueness and imperfections of this prior usage. Terms like force, field, heat, and so on used in modern physics have distinctly different meanings from earlier periods. The complex gene of the modern molecular biologist, with its flanking sequences, exons and introns, and other elaborations, is utterly different from the early "beads on a string" notion and even from the more sophisticated concept of H. J. Muller; yet the word "gene," which was first introduced by Johannsen in 1909, is still used to describe this entity. Because almost all scientific terms undergo a certain amount of change, it would be most confusing to introduce a new term with every minor change of meaning; new terms should be reserved for truly drastic changes. Indeed, technical terms must have a good deal of "openness" to permit the incorporation of further findings.

The second problem for the working scientist is that some terms have been unwittingly transferred from a given phenomenon or process to an entirely different one. This is well illustrated by T. H. Morgan's application of De Vries's term "mutation" to any sudden change in the genetic material; for De Vries, a mutation was an evolutionary change that would instantaneously make a new species. It was an evolutionary more than a genetic concept. It took the nongeneticists some 30 to 40 years to understand that Morgan's mutations were not the same as De Vries's mutations.[20] It is a basic principle of the language of science that a term which is in more or less universal use as the designation of a particular entity should not be transferred to a different entity. Violation of this principle invariably leads to confusion.

Perhaps most frequent and most confusing is the use of the same term for several different phenomena. In much of the philosophical literature, a great deal of logical sophistication is employed in the analysis of certain terms, but surprisingly little attention is paid to a term's possible basic heterogeneity.[21] Examples are "teleological," a term used for at least four entirely different processes; "group" (as in group selection), which again refers to four different kinds of phenomena; "evolution," which has been applied to three very different processes or concepts; and "Darwinism," a term which has continuously changed its meaning.[22]

Terminological ambiguity has had dire consequences from time to time in the history of biology. Darwin's failure to realize that the term "variety" was used differently by zoologists and botanists got him completely confused about the nature of species and of speciation.[23] A similar fate befell Gregor Mendel. He was uncertain about the nature of the kinds of peas he crossed and, like most plant breeders, he called heterozygotes "hybrids." When he tried to confirm the laws he had found by using "other" hybrids that were actually real species hybrids, he failed. The use of the same term "hybrid" for two entirely different biological phenomena thwarted his later research efforts.[24]

By far the most practical solution for such homonymy is the adoption of different terms for the different items. And whenever the possibility of confusing equivocation exists, precise definitions for each term in question should be proposed. If the designated concept or phenomenon changes its meaning, the definition should be revised appropriately. The definitions of most terms used in science are continuously modified as our knowledge increases. Just about every basic term in the physical sciences, for example, has been redefined again and again.[25]

Most philosophers seem to be quite reluctant to provide definitions, and perhaps this accounts for the many equivocations in the philosophical literature. The reason for this reluctance is that the term "definition" in the classical philosophical literature had a specific meaning that was a holdover from the scholastic tradition and was based on the principles of essentialism.[26] It seems that many philosophers use the term "explication" for that which a working scientist calls a definition.

To me the need for clear definitions is so obvious that I have never been able to understand why so many philosophers have been opposed to giving definitions. Popper, one of the most adamant opponents of definitions, revealed in his autobiography, *Unended Quest* (1974), why he held this view. He said he learned early in his youth that one should "never argue about words and their meanings, because such arguments are specious and insignificant." He was astonished to discover in his later readings "that the belief in the importance of the meanings of words, especially definitions, was almost universal." This, he says, was evidently the outcome of the power of essentialism. When Popper read Spinoza he found his writings "full of definitions which seem to me arbitrary, pointless, and question-begging." Popper here reveals what he was opposed to. It is the game of logicians to lay down definitions of words and then to operate in syllogisms with these.[27]

What Popper overlooked is that when a scientist demands a clear-cut definition, he is talking about something entirely different. What the scientist demands is an elimination of equivocation. If further scientific advance shows that the definition of a concept or process is incomplete or erroneous, the definition must be and will be changed. Without clear-cut definitions at all times, however, no progress in the clarification of concepts and theories is possible. It is my feeling as a practicing scientist that philosophers should give up their antipathy to definitions and should test by precise definitions whether or not the terms they use refer only to a single subject or to a heterogeneous mixture. This would put an end to a considerable number of controversies in the philosophical literature.[28]

Defining Facts, Theories, Laws, and Concepts

Quite a large philosophical discussion has developed around the meaning of terms such as hypothesis, conjecture, theory, fact, and law. For example, philosophers insist on making a distinction between a hypothesis and a theory, but I am unaware of a definition of theory that always permits such a sharp demarcation, especially in the life sciences. In any case, the scientist in the field or at the laboratory bench is usually not as precise in his use of these terms as the philosopher at his desk might wish. Whenever a scientist has a brainwave, he may say, "I just discovered (or invented) a new theory," when what he is actually describing might be considered by a philosopher a conjecture or hypothesis.

Another term that has become extremely popular in recent times is "model." To the best of my knowledge the term was not used once in the entire scientific literature on evolution or systematics prior to the last 20 years or so. How exactly does a model differ from a working hypothesis? Does a model have to be mathematical? How does it differ from an algorithm? I deliberately ask such "dumb" questions to indicate the need for more explanation from philosophers. All of these terms—conjecture, hypothesis, model, algorithm, theory—are sometimes used interchangeably by practicing scientists in formulating their explanations. (The reader is warned that I, too, often use the word "theory" in this looser sense.)

FACTS VERSUS THEORIES

A theory, to be sound, has to have a factual basis, but where does one draw the line between a theory and a fact? When does a universally supported and repeatedly verified theory come to be considered a fact? For instance, a modern evolutionist might say that the theory of evolution is now a fact. Strictly speaking, of course, a theory is never converted into a fact; rather, theory is replaced by fact. When the outer planets Uranus and Neptune showed irregularities in their orbits, the theory was advanced that there was a ninth planet, and in due time Pluto was indeed discovered. At that moment, the existence of Pluto was no longer a theory—it was now a fact. Similarly, after the structure of DNA was discovered and its control over protein synthesis was established, theories were proposed about a code that controls the correct translation of the information in the DNA. Rather quickly one of these theories proved to be the correct one, and the now-accepted genetic code was no longer considered a theory but simply a fact. In 1859 Darwin's ideas about the inconstancy of species and common descent were considered to be theories. The amount of evidence in favor of these "theories" and the absence of any counterevidence has, since then, led biologists to accept these theories as facts.

Facts, then, may be defined as empirical propositions (theories) that have been repeatedly confirmed and never refuted. Theories that have not yet been converted into or replaced by facts are nevertheless useful heuristic devices, particularly in areas of science where the sense organs are insufficient, such as in the microscopic and biochemical realm, or in sciences (such as cosmology and evolutionary biology) that construct historical narratives to explain past events.

UNIVERSAL LAWS IN THE PHYSICAL SCIENCES

What is the relationship of theories and facts to universal laws? Laws refer to processes with a predictable outcome, but many of the laws of physics, such as the law of gravity or the laws of thermodynamics, could just as well simply be called facts. That birds have feathers, although universally true, is simply a fact, not a law.

Those who have a high regard for natural laws are mostly thinking of the regularity of nature. Our human schedules are based on nature's regularities. We know in the summer that it will be followed by winter, and that each year trees will add a new ring of growth. Lyell's uniformitarianism was based on such observations. What happened in the past can be expected to happen today and in the future. When physicists wished to defend the certainty with which they held their theories, they would point out that theories in physics are based on universal laws that are without exceptions and are spatio-temporally unrestricted.

Regularities are abundant in the living world, too, but most of these regularities are not universal and without exception; they are probabilistic and very much restricted in space and time. Smart (1963), Beatty (1995), and other philosophers have maintained that there are few if any universal laws in biology. Of course at the molecular level many of the laws of chemistry and physics are equally valid for biological systems, and these are widespread in biology. But few, if any, regularities that have been observed in complex systems satisfy the rigorous definition of laws adopted by physicists and philosophers.

Most of the time, biologists who use the word "law" simply mean a logical general statement that is directly or indirectly open to observational confirmation or falsification, and that can be employed in explanations and predictions. Such "laws" are the basic constituents of any scientific analysis or explanation. But if one modifies the concept "law" to such an extent that it is applicable to any regularity or generalization in biology, then its usefulness in theory construction becomes rather questionable. Probabilistic theories, based on such so-called laws, rarely give the kind of certainty one is aiming for when using the word "law."

CONCEPTS IN THE LIFE SCIENCES

In biology, concepts play a far greater role in theory formation than do laws. The two major contributors to a new theory in the life sciences are the discovery of new facts (observations) and the development of new concepts. When one goes to a dictionary for the meaning of the term "concept," one gets a very broad definition. A concept may be any mental image. According to this definition, the number 3 when I think of it is a concept, and so is every other figure; every object of which I can form a mental image is a concept. But when a student of ideas speaks of concepts, he applies a much narrower definition, and yet there does not seem to be a good definition for "concept" in this narrower sense. Still, a biologist is virtually never in doubt as to what the important concepts of his field are. In evolutionary biology, for example, they include selection, female choice, territory, competition, altruism, biopopulation, and many others.

Concepts, of course, are not restricted to biology; they also occur in the physical sciences. What Gerald Holton (1973) calls themata is apparently what biologists refer to as concepts. I have the impression, however, that the number of basic concepts is rather limited in the physical sciences and in such fields of functional biology as physiology, where the discovery of new facts is very important. Indeed, some leaders in these fields have made statements indicating that they assume all progress in their science is due to the discovery of new facts. In most biological sciences, on the other hand, concepts play a large role. Not every new concept has as revolutionary an impact as did natural selection in evolutionary biology, but most recent advances in the more complex biological sciences (ecology, behavioral biology, evolutionary biology) are due to the proposal of new concepts.

The classical philosophy of science has made curiously little reference to the important role of concepts in theory formation. The longer I study theory formation, however, the more I am impressed by the fact that theories in the physical sciences are usually based on laws, those in biology on concepts. One can try to soften the seeming contrast by saying that concepts can be formulated as laws, and laws can be stated as concepts. But when the terms "law" and "concept" are rigorously defined, such a transformation is apt to run into difficulties. Here is a problem area which the philosophy of science, in its focus on physics, has rather neglected.

In the next chapter we will look more closely at unique factors biologists must take into account as they formulate and test their explanations of the living world.

CHAPTER FOUR

How Does Biology Explain
the Living World?

When a biologist tries to answer a question about a unique occurrence such as "Why are there no hummingbirds in the Old World?" or "Where did the species *Homo sapiens* originate?" he cannot rely on universal laws. The biologist has to study all the known facts relating to the particular problem, infer all sorts of consequences from the reconstructed constellations of factors, and then attempt to construct a scenario that would explain the observed facts of this particular case. In other words, he constructs a historical narrative.

Because this approach is so fundamentally different from the causal-law explanations, the classical philosophers of science—coming from logic, mathematics, or the physical sciences—considered it quite inadmissible. However, recent authors have vigorously refuted the narrowness of the classical view and have shown not only that the historical-narrative approach is valid but also that it is perhaps the only scientifically and philosophically valid approach in the explanation of unique occurrences.[1]

It is, of course, never possible to prove categorically that a historical narrative is "true." The more complex a system is with which a given science works, the more interactions there are within the system, and these interactions very often cannot be determined by observation but can only be inferred. The nature of such inference is likely to depend on the background and the previous experience of the interpreter; and therefore, not surprisingly, controversies over the "best" explanation frequently occur. Yet every narrative is open to falsification and can be tested again and again.

For instance, the demise of the dinosaurs was once attributed to the occurrence of a devastating disease to which they were particularly vulnerable, or to a drastic change of climate caused by geological events. Neither assumption was supported by credible evidence, however, and both ran into other difficulties. Yet, when in 1980 the asteroid theory was proposed by Walter Alvarez and, particularly, after the presumed impact crater was discovered in Yucatan, all previous theories were abandoned, since the new facts fit the scenario so well.

Among the sciences in which historical narratives play an important role are cosmogony (the study of the origin of the universe), geology, paleontology, phylogeny, biogeography, and other parts of evolutionary biology. All these fields are characterized by unique phenomena. Every living species is unique and so is, genetically speaking, every individual. But uniqueness is not limited to the world of life. Each of the nine planets of the solar system is unique. On earth, every river system and every mountain range has unique characteristics.

Unique phenomena have long frustrated the philosopher. Hume noted that "science cannot say anything satisfactory about the cause of any genuinely singular phenomenon." He was correct if he had in mind that unique events cannot be fully explained by causal laws. However, if we enlarge the methodology of science to include historical narratives, we can often explain unique events rather satisfactorily, and sometimes even make testable predictions.[2]

The reason why historical narratives have explanatory value is that earlier events in a historical sequence usually make a causal contribution to later events. For instance, the extinction of the dinosaurs at the end of the Cretaceous vacated a large number of ecological niches and thus set the stage for the spectacular radiation of the mammals during the Paleocene and Eocene, owing to their invasion of these vacant niches. The most important objective of a historical narrative is to discover causal factors that contributed to the occurrence of later events in a historical sequence. The establishment of historical narratives does not in the least mean the abandonment of causality, but it is a particularistic causality arrived at strictly empirically. It does not relate to any law but, rather, explains a simple, unique case.[3]

Causation in Biology

A scientific explanation is very often considered to be true if it is based on the discovery of the cause for an observed phenomenon, particularly of an unexpected phenomenon.[4] Causality in simple interactions is often highly predictive. In such cases—for instance, in certain chemical reactions—a definite cause can be designated with certainty. Most of the standard treatments of causality in the philosophical literature are based on problems in physics, where the effect of laws such as those of gravity and thermodynamics may give an unambiguous answer to the question "What is the cause of . . . ?"

However, such a simple solution is rarely available in biology, except at the cellular-molecular level. The problem is particularly perplexing whenever the effect is the end of a whole chain of events. It is perhaps a residue of teleological thinking that makes us search at the beginning of the process for the cause producing the predictable end effect. But in biology this approach is usually not successful; in fact, it is often misleading. It may be difficult, if not impossible, to pinpoint *the* cause in an interaction of complex systems, with the final effect being the last step in a long chain reaction. Here we may have to adopt a different way of thinking.

An interaction between two individuals, prior to its conclusion, goes through a whole series of stages, during most of which each of the acting individuals has several options available. Which of these he will choose is not strictly determined at the beginning of the stage but depends on a number of factors and contingencies. Strict causality can usually be construed only when the chosen option at each step of the chain of actions is looked at retrospectively. In fact, the whole process (even its random components) can be considered to have been causal when retrospectively considered. One could therefore say, somewhat paradoxically, that causation in complex situations is an *a posteriori* reconstruction, or, to put it differently, causation consists of a series of steps which, taken together, can be called the cause.

PROXIMATE AND ULTIMATE CAUSATIONS

There is a further complication as far as causation in biology is concerned. Every phenomenon or process in living organisms is the result of two separate causations, usually referred to as proximate (functional) causations and ultimate (evolutionary) causations. All the activities or processes involving instructions from a program are proximate causations. This means particularly the causation of physiological, developmental, and behavioral processes that are controlled by genetic and somatic programs. They are answers to "How?" questions. Ultimate or evolutionary causations are those that lead to the origin of new genetic programs or to the modification of existing ones—in other words, all causes leading to the changes that occur during the process of evolution. They are the past events or processes that changed the genotype. They cannot be investigated by the methods of chemistry or physics but must be reconstructed by historical inferences—by the testing of historical narratives. They are usually the answer to "Why?" questions.

It is nearly always possible to give both a proximate and an ultimate causation as the explanation for a given biological phenomenon. For instance, for the existence of sexual dimorphism one can give either a proximate physiological explanation (hormones, sex-controlling genes) or an evolutionary explanation (sexual selection, aspects of predator thwarting). Many famous controversies in the history of biology came about because one party considered only proximate causations and the other party considered only evolutionary ones. One of the special properties of the living world is that it has these two sets of causations. In the inanimate world, by contrast, there is only one set of causations—that provided by the natural laws (often combined with random processes).

PLURALISM

When one looks carefully at a biological problem, one can usually discover more than one causal explanation. Darwin, for instance (as we will see in Chapter 9), believed in both allopatric and sympatric speciation as explanations for the diversity of life, in natural selection and in inheritance of acquired characters as explanations for evolutionary change, in particulate inheritance (reversions) and in blending inheritance. Such pluralism of beliefs presents a problem for both verification and falsification. Producing evidence for natural selection would not necessarily falsify the inheritance of acquired characters, and falsifying the inheritance of acquired characters would not necessarily leave natural selection as the only other possible cause of evolutionary change.

Curiously, pluralism in biological explanation was much better appreciated by the old-time naturalists than by modern specialists. Biogeographers from Zimmermann on (in the eighteenth century) fully understood that discontinuities could be primary (dispersal jumps) or secondary (vicariance), but the present-day vicarianists not only act as if vicariance were the only possible solution but also act as if they had been the first to think of it! Some recent punctuated equilibrium enthusiasts write as though this was the only theory of evolutionary change possible, while earlier authors adopted plural solutions. Indeed, it is quite possible that in biology the majority of phenomena and processes must be explained by a plurality of theories. A philosophy of science that cannot cope with pluralism is not suitable for biology.

In biology a plurality of causal factors, combined with probabilism in the chain of events, often makes it very difficult, if not impossible, to determine *the* cause of a given phenomenon. For instance, the organisms found on a given island may have colonized it when it was connected to the mainland at an earlier period, or they may have arrived by overwater dispersal at a later period, or both. Any distributional discontinuity may be due to a secondary break of an originally continuous range (vicariance) or to dispersal across unsuitable terrain. A species may have become extinct owing to competition with another species, persecution by humans, a change of climate, an asteroid impact, or a combination of these. In many, perhaps most, instances it is not possible to determine with certainty which particular cause or combination of causes was responsible for a particular case of extinction in the geological past.

In almost all the classical controversies in biology, the opponents neglected to consider a third alternative to the two controversial viewpoints. For example, the reductionist explanations of the physicalists could not explain biological phenomena that have no equivalent in the limited inorganic realm, while the vitalistic counterproposals were equally deficient; organicism, a third viewpoint which combined the best of both, eventually prevailed (see Chapter 1). In the argument between chance and necessity, natural selection emerged as the third solution that ended the debate. And in the old preformation versus epigenesis argument, the solution to the controversy turned out to be the genetic program. Almost every protracted controversy in biology was terminated by the rejection of *both* previous explanations and the adoption of a new one.

PROBABILISM

In the days of strict physicalism, when everything was believed to be determined by an identifiable cause, to permit an outcome of a process to be also affected by chance or accident was considered unscientific. Therefore, Darwin's process of natural selection (which, though it did not proceed by chance, nevertheless assumed a good deal of randomness) was referred to by the physicist Herschel as the "law of the higgledy-piggledy." Actually, already in Laplace's day the role of stochastic (random) processes was appreciated by some scientists.

The reason why so many biological theories are probabilistic is that the outcome is simultaneously influenced by several factors, many of them random, and this multiple causation prevents any one factor from being 100 percent responsible for the outcome. If we say that a particular mutation is random, it does not mean that a mutation at that locus could be anything under the sun but merely that it is unrelated to any current needs of the organism or is not in any other way predictable.

CASE STUDIES IN BIOLOGICAL EXPLANATION

When philosophers of science discuss the formulation of scientific theories, almost all of the case studies cited deal with the physical sciences. Yet as we have seen, explanation in biology, and more particularly in evolutionary biology, may be rather different from that in the physical sciences. Thus, it might be helpful to examine a few cases that illustrate this difference more fully.[5]

Let me begin with the following simple situation. Members of the camel family are found in the living fauna only in Asia (and north Africa) and in South America. How can one explain such a discontinuous pattern of distribution? Louis Agassiz applied his theory of creation and simply postulated that God had created camelids twice, real camels in the old world and llamas in South America. When this suggestion became unacceptable after 1859, the hypothesis was proposed that camels must have existed in former times also in North America but then became extinct in that area. Paleontology has since confirmed this conjecture through finding a rich fossil camel fauna in North America.

A somewhat more difficult problem, of which Darwin was already aware, was the discontinuity of the fossil record. One of the more important components of Darwin's evolutionary paradigm was continuity. Evolution proceeds by gradual change. Yet, when one looked at living nature, all one saw was discontinuity. This was particularly conspicuous in the fossil record. New species, and more importantly, entirely new types of organisms, turned up quite suddenly in the fossil record, with no intermediates being found between them and their presumptive ancestors. To be sure, occasionally a "missing link" was found, such as *Archaeopteryx* between birds and reptiles, but even this fossil was separated by large gaps from its reptilian ancestors and from the true birds. Darwin stubbornly (and as we now believe, quite rightly) insisted that there must have been complete continuity, but that the fossil record was far too spotty to demonstrate this. His conclusion was not widely accepted for almost 100 years after the publication in 1859 of the *Origin*.

A contribution to a solution was provided in my 1954 paper on speciational evolution. I proposed that a peripherally isolated founder population could undertake a considerable ecological shift and genetic restructuring and become the ideal starting point for a new phylogenetic lineage. It is highly unlikely that such a small population would be preserved in the fossil record, however. This theory of geographic speciation was adopted and elaborated by Eldredge and Gould (1972) in their theory of punctuated equilibrium.[6] What we have here is a conspicuous conceptual shift from an essentialistic to a populational theory. Indeed, it is my impression that all more drastic theory shifts in biology are the result of a conceptual shift.

In many instances, a totally new causation may be postulated, while the bulk of the new theory remains remarkably similar to the old theory. For example, Darwin in 1839 explained the so-called "parallel roads" of Glen Roy in Scotland as old shorelines, ascribing their origin to a drastic elevation of the land. Having found marine shells at high altitudes in the Andes, having observed the dramatic rise of the Chilean coast after an earthquake, and making use of many other observations, Darwin did not consider such a major rise of that area in Scotland improbable, particularly since there was no other reasonable theory available. However, only a few years after Darwin's publication, Agassiz advanced his ice-age theory, and it became quite clear that the parallel roads were the shorelines of a glacial lake. Although Darwin himself later called his interpretation "a great failure," it was actually quite close to the correct solution. The essential insight was that the parallel roads were shorelines. Prior to the proclamation of the ice-age theory, the only way to explain such shorelines was to consider them ocean shores; furthermore, massive elevations of land were well established in the geological literature, particularly through the writings of Darwin's teacher, Charles Lyell. Explaining these same shorelines as due to glacial activities was not really a major change.

A similar situation pertains to that vast, and in many respects very magnificent, literature on design written by the natural theologians. It was possible to take over almost all of this literature into Darwinism simply by replacing the explanatory causal factor: it was not God who perfected the design but the action of natural selection. Scores of similar cases could surely be found in which the essential structure of a theory was left untouched; only the basic causal factor was replaced.

Cognitive Evolutionary Epistemology

All of epistemology is concerned with the problem of what we know and how we know it. In the last 25 years a movement called evolutionary epistemology (e.e.) has arisen which promotes a supposedly new way of looking at the acquisition of knowledge. One of its major representatives has referred to it in such extravagant terms as "a new Copernican revolution," while its opponents consider this claim to be misleading and the contributions of e.e. to be rather trivial.

The term e.e. has actually been applied to two entirely different processes which I shall call Darwinian evolutionary epistemology (analyzed in detail in Chapter 5) and cognitive evolutionary epistemology. Cognitive e.e. claims that certain "structures" in the brain, which evolved through a Darwinian selection process, permit humans to deal with the reality of the outside world, and that humans could not deal with their world if they did not have these brain structures. All individuals that were inferior in this capacity were sooner or later eliminated without leaving descendants.

Modern scientists fully understand that many perceptions of the "real world" are possible, and that our human senses provide only a very limited sampling of the characteristics of this world. The students of protozoans (beginning with Jennings) have revealed to us what the world is like to a one-celled creature. Von Uexküll has graphically described how different the world of a dog is from our world. We now realize that human beings see only the small slot of wavelengths represented by the colors from red to violet, from among a vast spectrum of electromagnetic waves. We do know of infrared rays manifested in warmth and of ultraviolet rays. We know that some flowers have ultraviolet coloration which is perceived by bees and other insects but not by us. Other animals can perceive and act upon magnetic information or hear above and below the range of sounds accessible to humans. We know that there is a vast olfactory world, much of it accessible to other mammals and certainly to insects, but not to us.

What determined the selection of those particular aspects of the total world that can be perceived by a human? The most plausible theory is that the ancestors of all organisms were able to survive and reproduce because they had the capacity to sense those aspects of their environment that are most important for their survival, and this, of course, is equally true for the human species. This thinking suggests that there are many "worlds," of which only one is accessible to us. That part of the world that is important for humans and their perceptions is sometimes referred to as the *middle world* (mesokosmos), the world of intermediate dimensions. It ranges from molecules to the Milky Way galaxy. Below it is the world of elementary particles, and beyond it is the transgalactic world of spacetime.

A solid table, the physicists remind us, is "in reality" not at all solid but consists of atomic nuclei and electrons that are far distant from one another. Most biologists I know accept the reality of this explanation and others (ranging from genes and quarks, to quasars, black holes, and dark matter, to the peculiar relations between the world of subatomic particles and the world of the ultragalactic cosmos). These phenomena cannot be perceived by human sense organs. The scientific realist, as people who hold this view are sometimes called, believes that the success of a theory warrants a belief in the existence of a postulated theoretical entity, and that such theoretical entities are as real as the observed ones. This scientific realism is shared by all the scientists I know.

But, frankly, in their everyday lives most people do not understand a table in this way, and this includes most physicists. Furthermore, no advance in our understanding of this smallest or largest world makes any contributions whatsoever to our understanding of the middle world, the "real world" as humans perceive it. Although instrumentation provided by physicists and engineers has opened up the fascinating subatomic world as well as the transgalactic one, none of these other worlds is part of our normal sensory world, and none of them contributes to our commonsense realism. And understanding them is not essential to our survival.

But how, then, is it possible that we can have ideas on such basic universal properties as time and space, if we cannot perceive them directly? Here the philosophy of Kant had a considerable impact on the thinking of some epistemologists. Kant, if I understand him correctly, believed that the brain is so structured that one is born with information about these properties of the universe. One must remember that Kant was an essentialist in much of his thinking and was convinced that the variable world of phenomena was represented in our thinking by one eidos for each class of variable phenomena which he called the *Ding an sich*. It existed *a priori*, that is, before any experience, hence prior to birth.

When Konrad Lorenz occupied Kant's chair in Königsberg in 1941, he developed a theory of evolutionary epistemology based on Kant's notion "that the perception and the thinking of man has functional structures which precede any individual experience." In order to be able to cope with the world, Lorenz said, a newborn must have various cognitive structures in his brain in just the same way that the newborn whale has fins for swimming. As our hominid ancestors shifted from one adaptive zone to another, appropriate mental structures were selected, by exactly the same process whereby structural adaptations were selected. These innate structures of our perception and thinking, says Lorenz, are the exact equivalent of morphological or any other kind of adaptations. It seems to me that Lorenz's suggestion is basically the same as the fact that eyes are laid down in the embryo long before they can be used for seeing.[7] Even the most primitive protists have an apparatus for sensing and responding to the dangers and opportunities they encounter in their habitat. More than a billion years of natural selection have elaborated the genetic program of the human species from that of a simple protozoan into that of mankind. Thus the new biological understanding of the nature of genetic programs has finally explained what for such a long time had been a great mystery for the philosophers.

I believe that one must accept the idea that during the evolution of humans from primates, a brain rapidly evolved that was able to solve problems considerably beyond the capacity even of a chimpanzee. But this still leaves unanswered the question: "How specific is the structuring of the modern human brain?"

CLOSED AND OPEN PROGRAMS

There is much to indicate that physically the human brain reached its present capacity nearly 100,000 years ago, at a time when our ancestors were culturally still at a very primitive level (see Chapter 11). The brain of 100,000 years ago is the same brain that is now able to design computers. The highly specialized mental activities we see in humans today seem not to require an ad hoc selected brain structure. All the achievements of the human intellect were reached with brains not specifically selected for these tasks by the Darwinian process.

To be sure, different human capacities are controlled by different areas in the brain. But in view of our present great ignorance about the workings of the human brain, it would be misleading to become too specific at this time in our speculations about the brain structures that permit human cognition and recognition of the world. Yet based on what we know at this moment, it appears that one might recognize three kinds of areas in the brain.

First, the brain seems to contain areas that from the very beginning are rigidly programmed. Instincts in the lower animals, and reflexes and most locomotory patterns in both lower and higher animals, are examples of these "closed programs." But whether more complex behaviors of the human species (and if so which) belong to this category is unknown. Research in the area of infant behavior and temperament indicates that there may be more rigidly programmed behaviors than we used to think.[8]

The brain also seems to contain areas that are suitable for "open programs." This information is not rigidly programmed in the way that instincts are, but specific areas in the brain are set aside to accept such information if it is available in the environment of the young organism. Many components of our cognitive equipment, such as the capacity to learn languages or to adopt ethical norms, are apparently best acquired at certain early ages and are not easily displaced or forgotten once acquired. These categories of learning seem to have much in common with the simple "imprinting" of the ethologists. The young gosling, during an early sensitive period, becomes "imprinted" on the gestalt of its mother. This "object-to-be-followed" is inserted in the gosling's brain in an area evidently ready for the acceptance of this information. Similarly, every new experience of a developing human is recorded in the appropriate brain space and reinforces associated experiences that had previously been recorded by the brain.[9] The components of our knowledge of the world with which we are born, as described by Kant as well as Lorenz and other evolutionary epistemologists, are perhaps best understood as open programs.

NB

Finally, the brain seems to contain generalized areas which permit the storage (memory) of all sorts of information acquired throughout the course of life. At the present time we know virtually nothing concerning a possible subdivision of the brain for different categories of such general information. Short-term and long-term memory may be examples of these subdivisions.

Cognitive evolutionary epistemology is particularly concerned with the second class in this list. It deals with brain areas that evolved through selection to supply the newborn with suitable open programs in which to store important and specific cognitive information. There is nothing metaphysical or essentialistic about such brain areas; they are simply a product of Darwinian evolution. What is still largely unknown is the degree of specificity of these areas. It would seem probable that much of the specificity is acquired after birth. This is indicated by the relative ease with which, in a young person, many functions of large destroyed parts of the brain can be taken over by other areas.

How does all this add up with respect to an evaluation of cognitive e.e.? I conclude that highly specific brain structures are not needed for the perception and understanding of our world. On the whole, it would seem that the evolutionary improvement of the central nervous system does not necessarily lead to highly specific neural structures but rather to a continuously improved general structure of the brain. As a result, not only is it able to cope with the actual challenges that faced primitive humans but it also has capabilities, such as those required for playing chess, that were not being called upon at the time when these improvements of the brain were selected. As a whole, it appears to me that cognitive e.e. is nothing revolutionary but a natural outgrowth of applying Darwinian evolutionary thought to neurology and epistemology.

The Quest for Certainty

The aim of science is often described as the search for truth, but what is truth? Darwin's Christian opponents never questioned the truth of every word in the Bible, leading them to the conclusion that everything in this world had been created by God. What in former eras were daring unorthodoxies, such as that the earth moves around the sun, are now considered absolute truths. That the earth is round and not flat (as was previously believed) is no longer denied by any reasonable person. The historian of science knows how many "unquestioned truths" of former periods have subsequently been shown to be errors. Prior to Kepler, astronomers took it for granted that the orbits of all heavenly bodies are perfect circles. Prior to Darwin, most philosophers were sure that species are constant. Until the 1880s, it was universally accepted that characteristics acquired during one's lifetime could be passed on to one's offspring. None of us knows what silent assumptions our generation makes that will ultimately be refuted by further scientific advances.

That the sequence of fossils in the earth's strata documents evolution is now accepted by scientists as an irrefutable truth. But many other findings of science are still tentative. They may have a high degree of certainty, but we would not be greatly disturbed by their eventual replacement by an either slightly or drastically revised alternate theory. Scientists no longer insist on "absolute truth." They are satisfied if a particular theory has withstood all attempts at falsification and if it explains everything that it is supposed to explain. For centuries it was believed that Newton's equations were the ultimate truth. Eventually, however, it was shown by Einstein's relativity theories that under certain conditions these equations are not correct, no matter how adequate they are in the normal terrestrial situation.

The commonsense consensus would seem to be that most conclusions of science are so well established that they can be considered to be certainties, while others are only provisional truths with varying degrees of certainty. If there is competition between two theories and it cannot be clearly established which of the two is "more true," Laudan (1977) suggests adopting that theory which is more successful in solving problems, or which has solved the most problems.

The truth of explanations, however, is often vulnerable. That birds acquired their feathers assisted by natural selection is a proposition that is almost surely true, but like most things that happened in the distant past, it can probably never be established unequivocally—that is, it cannot be proven. *Why* the acquisition of feathers was of selective advantage is even more difficult to prove: Was it for protection against cold in these warm-blooded vertebrates, or for protection against excessive solar radiation?[10]

There are observations in every branch of science that are still totally unexplained. Why has the phenotype of certain invertebrates (particularly so-called living fossils) remained virtually unchanged for more than a hundred million years, while their associates in the same faunas either have become extinct or evolved drastically? Why do two kinds of birds seem equally successful, one in which the male actively takes part in the raising of the young and the other where the male does not? (The answer may be what the young are fed, insects or fruit.) The number of such puzzles was far greater 50 or 100 years ago, and in the meantime a remarkably high percentage of such cases has been explained satisfactorily—for instance, why members of the sterile cast of social insects participate with such devotion in the raising of the offspring of the queen.[11] Biochemistry has been able to elucidate the nature of almost all physiological puzzles. The most important remaining puzzles concern the explanation of the most complex processes in organic life, the development of the fertilized egg up to the adult stage and the functioning of central nervous systems. Most individual processes in these two important fields are already reasonably well understood, but the explanation of the integration of the individual processes and their control is still a little beyond our comprehension.

In the light of these remaining uncertainties, some nonscientists have gone to the extreme of claiming that *nothing* found by science has any degree of certainty. And even some philosophers have questioned whether we can ever find the ultimate truth about anything. This uncertainty has led to the question which we will consider in Chapter 5: "Does science advance?"

CHAPTER FIVE

Does Science Advance?

Practically all working scientists, and indeed most lay people with an interest in science, are convinced that we are making steady advances in our understanding of nature, as successive generations of scientists fill in more and more parts of the "true" story of how the world works. According to this view, there may be some questions that we will never be able to answer ("Why is there our world?" "Why is it constructed as it is?"), but in every branch of science a vast number of questions can still be identified that would seem to be accessible to further research.

This conviction that science has advanced, and will continue to advance, is by no means shared by everybody, however. During the last 50 years the shift in the philosophy of science from strict deter-minism and a belief in absolute truth to a position in which only an *approach* to truth (or presumed truth) is recognized has been inter-preted by some commentators as evidence that science does not advance. This has led the antiscience movement to argue that science is a wasteful activity because it does not lead to any final truth about the world around us.

When one reads the current biological literature, one can perhaps understand how such a negative view could have arisen. To outside observers, the seemingly unresolved controversies surrounding punc-tuated equilibria, the role of competition in ecosystems and of dispersal in biogeography, the control of biological diversity, the adaptationist program, and the definition of species (to mention only a few of the issues discussed in the chapters that follow) might easily lead to the conclusion that no consensus is in sight, hence no hope for any real progress. Even a handful of scientists themselves believe that we may be reaching the limit of the questions that can be answered by science.[1]

Throughout the philosophy of science, one finds widespread objection to the notion of scientific progress, which Kitcher (1993) has referred to as "the Legend." According to the Legend, science has been very successful in attaining "the goals of science . . . Successive generations of scientists have filled in more and more parts of the complete true story of the world . . . Champions of Legend . . . saw an overall trend toward . . . a better and better approximation to truth." My confession that I adhere to the Legend will undoubtedly make these critics consider me old-fashioned. But what I would like to know is, to what science do these critics refer? I must admit that developments in the sciences I know best fit the Legend remarkably well.

For example, the history of geology from Werner and Lyell up to modern plate tectonics, taken together with the history of organic evolution from Lamarck to the evolutionary synthesis of the 1940s, must surely be considered as progress over the previous belief in an unchanging world. The progression from Ptolemy to Copernicus, Kepler, Newton, and modern astrophysics is a story of continuous improvement in our understanding of the cosmos. Changes in scientific thinking from Aristotle to Galileo, Einstein, and quantum mechanics is another saga of steady advance.

Similar series of progressive stages can be cited for morphology, physiology, systematics, behavioral biology, and ecology. The development of molecular biology since the 1940s has been an uninterrupted run of achievements. Where there was virtually nothing prior to the 1940s, we now have a well-established megascience. All the major advances in medicine rest on advances in biology or other basic sciences. I could take up one problem in biology after the other and show how successive theories have become increasingly more powerful in explaining the known facts.

But what exactly do we mean by the terms "scientific advance" or "scientific progress"? We mean by them the establishment of scientific theories that explain more and better than earlier ones and are less vulnerable to refutation. In most sciences, better theories permit better predictions, and they are less likely to be replaced by other conjectures. Which of two or more theories is the "better" one is often exactly the point of a scientific controversy. The history of science shows, however, that in due time the controversies concerning a particular problem are somehow resolved, and eventually one theory is generally acknowledged to be better than its competitors. The resolution of many historical controversies was achieved by the rejection of *both* opposing theories and their replacement by a third one.

Quite often, a theory becomes so successful that in the end it no longer has any competitors. Yet the fact that a particular theory at a certain time is the only one explaining a process or phenomenon does not necessarily mean that it is the final word. The large number of theories that were once universally accepted but were subsequently so thoroughly refuted that they are now unanimously considered invalid is further evidence of scientific progress. A few of the best known among literally hundreds of such theories might be mentioned: Schwann's theory of the origin of new cells from the nucleus, blending inheritance, the quinarian relationship of taxa, the inheritance of acquired characters, and untold theories in physiology. These now-refuted theories, when they were first proposed, were usually the best possible explanation at that time, based on the then-existing information and conceptual framework of the field. But scientists are rarely satisfied with any theory; they always attempt to improve it or replace it by a better or more comprehensive one. The theories that have taken their place have withstood numerous attempts at refutation and are consistent with the available evidence up to the present time.

Some authors, among whom Charles Darwin is perhaps foremost, have a remarkably high batting average for the success of their theories. But even Darwin proposed theories that have since been refuted. Among them are pangenesis and sympatric speciation owing to the principle of divergence. The history of genetics provides particularly good illustrations for the conclusion that much advance in science consists of the refutation of erroneous theories.

To be sure, not every theory change in science is necessarily evidence of progress. Indeed, when in the late 1890s the theory that "nuclein" was the genetic material was abandoned in favor of proteins, it was later discovered to have been a backward step. The same can be said for the typological-saltational evolutionary theories of the Mendelians (Bateson, DeVries), who rejected the prevailing Darwinian concept of gradual populational evolution. The historiography of biology abounds in examples of such temporary retrograde developments. What these cases have taught us is that it is an error to totally abandon a seemingly refuted theory until it has been exhaustively tested and found to be unquestionably erroneous.

The pathway to new insights is by no means necessarily rectilinear. Indeed it often is a "zeroing-in," a zigzag approach making use of the principle of reciprocal illumination. Every solution of a scientific question, large or small, leads to new questions; there is usually an unexplained residue, the so-called black boxes—somewhat arbitrary assumptions that are still in need of fuller analysis and explanation. In that sense, there will never be an end to science.

Not all activities that engage scientists' time and attention necessarily lead to scientific advance. In every field there are clerical minds who enjoy preparing lists and other compilations, who like to establish data banks and become occupied by other activities that will be helpful to other workers but do not noticeably advance the field. Most workers—perhaps for good reasons—are afraid to look at the great unsolved problems of their fields. Instead they essentially duplicate what has already been done by others. For instance, they will study in *Drosophila virilis* what has already been established in *Drosophila melanogaster*. Others produce a rich body of new facts but fail to develop any generalizations from these facts.

Some workers restrict themselves to a highly specialized problem and fail to establish intellectual and particularly conceptual contact with workers in neighboring fields. Scientific explanations often make use of information and concepts from a number of adjacent fields, and a theoretical advance in one field may often have repercussions in several related fields. Sometimes progress in science is expressed not by simply refuting another theory but by broadening the explanatory basis which unites or synthesizes several scientific disciplines.

Most of those who have attacked the notion of scientific progress have been philosophers or other nonscientists who simply do not have the expertise to be able to evaluate whether or not there has been any real progress in our understanding. Everything I know about science leads me to disagree with the claims of these critics. Most principles and theories of current science have remained unshaken for 30, 50, 100, some more than 200 years. Our basic understanding of the world is now remarkably robust.

There are a few major exceptions, such as our understanding of the brain and of the cohesion of the genotype, but it must be emphasized these are exceptions. Yet skepticism about scientific advance is still sufficiently widespread outside of science to justify further documentation for steady progress in various fields of science, particularly biology. In order to substantiate the claim for real progress, I shall analyze in detail a concrete case study.

Scientific Advance in Cell Biology

Cytology—the scientific study of cells—is particularly suitable for this purpose.[2] This field was made possible by the invention of the microscope. The first work in cytology was published in 1667 by Robert Hooke under the title *Micrographia,* in which the word "cell" was used for the first time. Although many microscopic objects were described in the ensuing 150 years by three outstanding microscopists, Grew, Malpighi, and Leeuwenhoek, the study of microscopic objects was more an amusing entertainment than serious science. Little that was new was described from 1740 to 1820. Although cells were occasionally referred to, the references seemed to emphasize fibers and other longitudinal structures rather than cells.

The major advances between 1820 and about 1880 or 1890 were made possible by technical improvements in the lenses (the most important of which were made by Abbe) and by the discovery of oil immersion. The lighting of objects was also constantly improved as well as the methods of fixing tissues and any living material, and, finally, by the use of all sorts of dyes to produce contrast between cell wall, cytoplasm, nucleus, and cellular organelles. Some of the most important early discoveries by investigators such as Brown, Schleiden, and Schwann were made with remarkably primitive self-made micro-scopes. In the early nineteenth century, however, a number of optical firms began to produce ever-improved microscopes, and this greatly facilitated the study of cells and helped to popularize cytology. The inadequacy of the early instruments often led to erroneous observa-tions, and was one of the reasons for some of the early controversies in cytology.

One gains the impression from most histories of biology that the study of cells began with Schleiden and Schwann. However, F. J. F. Meyen (1804–1840) published even earlier a remarkably accurate and well-informed monograph on plant cells.[3] He described the multipli-cation of cells by division, used iodine to stain starch inclusions in plant cells, and gave an exact description of chloroplasts. If he had not died so young, no doubt his would have become an honored name in the history of biology. But Meyen was not alone; there were at that period about half a dozen other investigators who made substantial contributions to the accurate description of cells.

In November 1831 Robert Brown announced his discovery of a body in all cells which he called the nucleus. But he refrained from speculating on its significance. This was done by M. J. Schleiden in a paper published in 1838 in which he claimed that new cells originate by the growth of the nucleus. He therefore renamed it the cytoblast. The nucleus itself, he said, was formed *de novo* from the liquid of the cell content. Quite evidently this was an epigenetic theory of the origin of cells, fitting into an intellectual environment in which any kind of preformation was frowned upon. Nevertheless, Meyen immediately published a rejoinder to Schleiden in which he reiterated his obser-vation of the formation of new cells by the division of old cells, a process that for Schleiden presumably smacked of preformation. It did not help Meyen's thesis that he held a number of other ideas about the cell nucleus that were rather erroneous.

Schleiden, a botanist, had done his cytological investigations on plant cells with their well-formed cell walls. He confirmed a conclusion that Meyen had essentially already reached, which was that a plant consists of nothing but cells, even though some of them are greatly modified. But what about animals? Do they also consist of cells? This was demonstrated in 1839 by Theodor Schwann, who was able to show for one animal tissue after the other that the components of these tissues, no matter how different they seemed from one another, were nothing but modified cells. Schwann, however, also confirmed, in a very detailed investigation, Schleiden's erroneous theory that new cells originate from nuclei. He only added another process that new nuclei could originate from unformed intercellular material.

Few publications in biology have ever caused such a sensation as Schwann's magnificent monograph. It demonstrated that animals and plants consist of the same building blocks—cells—and that a unity therefore exists throughout the entire organic world. Furthermore, the cellular composition of both animals and plants demonstrates that cells are the elementary components of organisms. It was a vigorous endorsement of reductionist thinking.

Schleiden later published a detailed presentation of his theory of science with strong emphasis on induction and a severe critique of the then-fashionable theories of science of Schelling and Hegel. Yet, it is quite clear that Schleiden was not nearly as inductive and empirical as he thought he was, and his ultimate conclusions were all teleological. He clearly based his theory of science on Kant by way of Fries. A similar teleological worldview was true for Schwann, who was a devout Catholic.

The Schleiden-Schwann theory of the origin of new nuclei from the cytoplasm or other unformed organic substances fit well not only into the epigenetic thinking of the embryologists but also into the theory of spontaneous generation, which was still widely accepted at that time. It is another illustration of the influence of ideologies on the acceptability of theories. The theory of the possibility of a free formation of new nuclei and cells in unformed organic material was thoroughly refuted by Robert Remak in 1852. He showed that in a developing frog embryo, beginning with the first cleavage division, every cell of every tissue was the result of the division of a preexisting cell. In 1855 he followed this up with a larger, well-illustrated monograph, in which the Schleiden-Schwann theory was even more thoroughly refuted. In the same year Virchow adopted Remak's conclusions and coined the famous motto *omnis cellula e cellula* ("all cells from

cells"). Not surprisingly, Virchow was also a determined opponent of
the theory of spontaneous generation.

It is not altogether easy to determine what really caused the change
in the theory of the origin of cells. Presumably the improvement of
microscopes and microscopic techniques was involved, as well as the
choice by Remak of a particularly suitable material, the developing
frog embryo. On the other hand, the new theory was in seeming
opposition to epigenesis and to the theory of spontaneous generation,
both still prevalent at that time. It seems, at least in this case, that the
empirical findings simply swept away any misgivings over the seeming
violation of broadly held ideas.

UNDERSTANDING THE NUCLEUS

The new cell theory originally had no use for the nucleus, even though
Remak had shown clearly that a division of the nucleus preceded the
division of the cell; this observation was categorically denied by others,
including the otherwise so pioneering Hofmeister. As a result it took
another 30 years before the slogan *omnis nucleus e nucleo* ("all nuclei
from nuclei") could be coined by Flemming.

It was really the process of fertilization that ultimately provided the
most important clues. It started with the proof supplied by Kölliker
(for the egg) and by Gegenbaur (for the spermatozoon) that these two
reproductive elements are cells. What role they played in fertilization
and development was, however, in the beginning very controversial.
For the physicalists fertilization was nothing but a physical phenomenon, consisting of a transmission of excitation produced by the contact
of the spermatozoon with the egg cell. Fertilization for them was
simply the signal that initiated the cleavage of the egg cell. For their
opponents it was the "message" which the spermatozoon brought to
the egg that was the truly significant aspect of fertilization.

Before the latter view could eventually achieve victory, a number of
erroneous ideas about development had to be eliminated. Most important among these was preformation, the belief that a miniature
organism was encapsulated in either the egg or the spermatozoon.
Beginning with Blumenbach, this idea was ridiculed so unmercifully

that it eventually was replaced by the theory of epigenesis, the belief that development started from an entirely unformed mass which was given form by some extraneous force.

The second idea that had to be accepted was that of the equal contribution of egg and spermatozoon to the characteristics of the developing embryo, in other words, a consideration of the genetic aspects of fertilization. This proof was first provided by Koelreuter, who in the 1760s proved this point conclusively in his hybridization experiments. Although Koelreuter's work was widely ignored, findings similar to his were made in subsequent years by many others, and the idea that the spermatozoon played a much more important role than merely initiating the cleavage of the fertilized egg was eventually accepted. Amazingly, as late as the 1870s, Miescher, the discoverer of nucleic acid, still adhered to the physicalist interpretation.[4]

The entry of the spermatozoon into the egg, and sometimes even the fusion of the male nucleus with the egg nucleus, had been observed repeatedly between the 1850s and 1876, but these observations were misinterpreted owing to the erroneous conceptual framework of the investigator. It was Oskar Hertwig (1876) who clearly showed that fertilization consisted of the penetration of a spermatozoon into the egg, that the spermatozoon provided a male nucleus which fused with the egg nucleus, and that the development of the embryo was initiated by the division of the newly formed nucleus of the zygote that had been formed by the fusion of the male and egg nucleus. These observations were fully confirmed and expanded by H. Fol in 1879.

The idea—widespread in the preceding decades—that the cell nucleus is dissolved prior to every division of a cell was now clearly refuted, at least for the process of fertilization, and improved microscopic techniques soon demonstrated that every cell division was initiated by the mitosis of the cell nucleus.

What was not fully understood at the time was that fertilization by the spermatozoon played a dual role. It imported the genetic material of the father into the egg but it also gave the signal for the beginning of the development of the zygote. That these are two entirely different roles was not understood by the physicalists. When Loeb was able by chemical means to initiate the development of unfertilized eggs, he made claims about such artificial parthenogenesis which showed that he was entirely unaware of the genetic role of fertilization.

By the 1870s it had become quite clear to the foremost workers in the field that the fusion of the spermatozoon nucleus and the egg nucleus had a genetic significance. Just what this significance was and how the two nuclei could transmit the genetic properties of the parents was still totally obscure. What was next required was the discovery and correct description of the reduction division during the meiosis of the maturing germ cells and the appreciation that the essential component of the nucleus were the chromosomes; this was accomplished by Weismann, van Beneden, and Boveri.

The empiricists, those who did the superb microscopic work, often missed the correct interpretation of their findings simply because they did not have an appropriate theoretical framework. Often they did not ask the question why something was happening. Here Roux was exemplary. He asked very perceptively: Why is the complex process of mitosis necessary? This process is so wasteful in time and so seemingly unnecessarily complex. Why not simply halve the nucleus and give one half to one daughter cell and the other to the other daughter cell? Roux quite correctly concluded that the elaborateness of the process of mitosis is justifiable only if the nuclear material is qualitatively highly heterogeneous and a method must be used to make sure that each daughter nucleus receives its share of each of the qualitatively different components of the original nucleus.

What is equally interesting about this period is that many correct observations and theories were subsequently ignored, only to be rediscovered later. Perhaps I should say, "with the true significance" being discovered only later. For instance, Roux more or less abandoned his own valid theory of mitosis because it seemed to be contradicted by some of the observations on developing eggs. And van Beneden's completely correct observation that the chromosomes of the sperm nucleus did not fuse with those of the egg nucleus, providing the observational basis for Mendel's findings, was largely ignored until after 1900.

None of the speculations on theory formation found in the literature

of the philosophy of science is valid for the highly involved developments of this period, including wrong observations and false guesses.
Advances were sometimes made by new discoveries, sometimes by new
theorizing. It was sometimes material from a new organism that
permitted advances, like Oskar Hertwig's sea urchin egg and Remak's
frog embryo, and sometimes new technologies, like the aniline stains
so successfully used by the later cytologists. The one thing that is
perfectly evident is that what was needed was an abundance of new
observations and new theories on which a Darwinian selection process
could operate. Sooner or later, one particular observation or interpretation would turn out to be unassailable and be accepted as "true."
Even it may be ultimately refuted, just as the assumption that proteins
are the material of inheritance was eventually refuted, though it had
been accepted for 30 or 40 years as more or less the truth. The protein
hypothesis was so firmly established that when it was finally replaced
by the DNA hypothesis, some leading investigators, such as Goldschmidt, still refused to believe it.

During the 40 years after 1880, improvements in microscopy permitted ever more accurate descriptions of nuclei and their changes in
the mitotic and meiotic cycles and an explanation of the meanings of
these changes. The acquisition of this understanding is a highly complex story, with contributions made both by superb technicians, who
supplied excellent descriptions of the various aspects of the processes
of maturation and fertilization, and by brilliant theoreticians.[5]

UNDERSTANDING THE CHROMOSOMES

The starting point of the ensuing speculations was the observation
that well-formed chromatin bodies, later called chromosomes, are
observed during cell division (mitosis) but were seemingly replaced
by a granular mass or a network of thin threads in the resting stage
of the nucleus. The problem was to find a meaning for what happens
when this irregular chromatic material is converted into the well-
defined chromosomes, particularly after it had been shown that each
species had a fixed number of mitotic chromosomes. It was rather
difficult at first to develop a theory as long as one had no idea
whatsoever as to what the biological role of the chromatin was. Al-

though it had been stated early on that chromatin was nothing but nuclein, this conclusion was by no means generally accepted, and since no one really knew what the function of nuclein was, this more precise identification was of no help either.

At this stage it was Weismann who insisted that the genetic material was located in the chromosomes, and although the details of his theory of inheritance were quite wrong, it steered attention in the right direction. The person who contributed the most to an understanding of the chromosomes was Boveri. He started with the simple observation that there was a fixed number of chromosomes during mitosis, and in favorable material he could demonstrate the individuality of these chromosomes, that is, he could recognize that each of the chromosomes had certain diagnostic characteristics. After these chromosomes had been "dissolved" into the nuclear matter of the resting nucleus, Boveri could demonstrate that during the next mitotic cycle the very same number of chromosomes re-formed as in the last mitotic cycle, and furthermore that they had the same individual characteristics as in the previous cycle. This led him to the continuity theory, according to which the chromosomes never lose their identity during the resting phase of the nucleus, but continue throughout the life of the cell. Although this theory was heavily attacked by other leading cytologists, including Hertwig, it eventually became the basis of the Sutton-Boveri chromosome theory of inheritance.

Boveri's theory was based on inference. The continuity of the chromosomes could not be observed directly. Was there some underlying deeper concept or ideology that gave Boveri the strength of conviction that he was right? Did his opponents have some other underlying concept or ideology which induced them to be sure that Boveri was quite wrong? Unfortunately I have been unable, on the basis of the existing literature, to come to a conclusion on this question. I rather suspect, however, that there must have been something in the conceptual background of Boveri and Hertwig that led to their drastic difference of opinion. Needless to say, neither of them invoked any laws to back up their opinions. Their conclusions were based on observations and what each of them thought was a logical inference from these observations. Up to now this disagreement has not yet been explained in any terms that would shed light on the philosophers' controversies on theory formation. Was perhaps the argument about the continuity of the chromosomes through the resting phase of the nucleus still a remnant of the preformation-epigenesis controversy, with Hertwig the epigenesist and Boveri the preformationist?

There was no slowing down in the advances of the understanding of the cell after 1900. At first the major contributions were made by genetics and cell physiology, subsequently followed by the exploration of the fine structure of the cell with the help of electron microscopy, and finally the study of all components of the cytoplasm by molecular biology. Even though observations were almost invariably the starting point of new developments, theory formation clearly was not the result of simple induction. Rather the observations raised puzzling questions, which led to conjectures that were either falsified or confirmed, ultimately resulting in new theories and explanations.

The history of cytology illustrates in the most graphic manner the gradual progress of science, the failure of erroneous theories, the struggle among competing theories, and the ultimate victory of the interpretation that at present has the greatest explanatory value. And it is indisputable that the interpretation of the cell and its components which is now adopted is infinitely superior to the concept of the cell prevalent 150 years ago.

Does Science Advance through Revolutions?

If we can conclude from this and other case studies that science makes steady advances in our understanding of nature, the question must next be asked how these advances occur. This highly controversial issue occupies a large portion of the literature of the contemporary philosophy of science. One can distinguish two major schools: (1) Thomas S. Kuhn's theory of scientific revolutions versus normal science, and (2) Darwinian evolutionary epistemology.

Few publications in the philosophy of science have created as great a stir as did Kuhn's *Structure of Scientific Revolutions* in 1962. According to Kuhn's original thesis in the first edition, science advances through occasional scientific revolutions, separated by long periods of "normal science." During a scientific revolution, a discipline adopts an entirely new "paradigm," which then dominates the ensuing period of normal science.

Revolutions (paradigm shifts) and periods of normal science are only two aspects of Kuhn's theory. Another one is a supposed incommensurability between the old and the new paradigm. One of Kuhn's critics has claimed that Kuhn used the term paradigm in at least twenty different ways in the first edition of his book. For the most important of these concepts, Kuhn later introduced the term "disciplinary matrix." A disciplinary matrix (paradigm) is more than a new theory; it is, according to Kuhn, a system of beliefs, values, and symbolic generalizations. There is a considerable similarity between Kuhn's disciplinary matrix and terms such as "research tradition" used by other philosophers.[6]

Many authors were able to confirm Kuhn's conclusions; perhaps many more were unable to do so. The numerous aspects of his thesis cannot be discussed profitably without looking at concrete cases and asking whether theory change did or did not follow Kuhn's generalizations. I have therefore analyzed a number of major theory changes in biology with this question in mind.

PROGRESS IN SYSTEMATICS

In the science of animal and plant classification (systematics; see Chapter 7), we can distinguish an early period, from the herbalists in the sixteenth century to Linnaeus, during which most classifications were constructed by logical division, and the nature of the changes made from one classification to another depended on the number of classified species and on the weighting of different kinds of characters. This type of methodology is referred to as *downward classification.*

In due time it was realized that downward classification was really a method of identification, and it was supplemented by a very different method, *upward classification,* consisting of the arrangement in a hierarchical fashion of ever-larger groups of related species into classes. However, the method of downward classification continued to exist side by side, being used in keys in all taxonomic revisions and monographs and in field identification guides. Upward classification was first used by some herbalists, later by Magnol (1689) and Adanson (1763), but this method did not begin to be generally adopted until the last quarter of the eighteenth century. There was no revolutionary replacement of one paradigm by another one, since both continued to exist, although now with different objectives.

One would have expected that the adoption of Darwin's theory of common descent in 1859 would have produced a major revolution in taxonomy, but this was not the case. In upward classification, groups are recognized on the basis of the greatest number of shared characters. Not surprisingly, the taxa thus delimited consisted usually of descendants of the nearest common ancestor. Hence, Darwin's theory supplied the justification for the method of upward classification, but it did not result in a scientific revolution in systematics.

One hundred years later, after 1950, two new schools of macrotaxonomy were established, numerical phenetics and cladistics. Did these amount to revolutions? Phenetics produced rather unsatisfactory classifications so that it did not have a major impact. Furthermore, it supplied a new methodology but not really a new concept. By contrast, if one looks at the volume of the current literature, one might be inclined to say that cladistics indeed produced a major revolution. Actually, the approach of recognizing taxa by jointly derived characters was already previously widely practiced, as Hennig himself points out (1950). Yet, it is evident that the vigorous and consistent application of cladistic analysis unquestionably had a major impact.

Even if one were to designate this a scientific revolution, however, it did not proceed as described by Kuhn. There was no sudden replacement of one paradigm by a different one, because two systems were existing side by side: Hennig's ordering system (cladification) and the traditional Darwinian methodology (evolutionary classification). They differed not only in methodology but also in objective. The cladistic system was interested only in discovering and representing phylogeny, while the endeavor of the evolutionary system was to construct taxa of the most similar and most closely related species, an approach particularly useful in ecology and life history studies. The two approaches can continue to exist side by side, since they differ entirely in their objectives.

PROGRESS IN EVOLUTIONARY BIOLOGY

Evolutionary biology provides another testing ground for the theory of scientific revolution. The simple picture of the biblical creation story began to lose credence by the end of the seventeenth century. In the eighteenth century, when the long duration of geological and astronomical time was beginning to be appreciated, when the bio-geographic differences of the different parts of the world were estab-lished, and when an abundance of fossils were described, various new scenarios were proposed, including repeated creations, all of them, however, operating with new origins. These new theories existed side by side with the biblical story of creation, which was still supported by the vast majority. The first to seriously undermine these views was Buffon, many of whose ideas were in complete opposition to the essentialistic-creationist world picture of his time. Indeed, it was from his ideas that the evolutionary thinking of Diderot, Blumenbach, Herder, Lamarck, and others was derived. When in 1800 Lamarck proposed the first theory of genuine gradual evolution, he made few converts; he did not start a scientific revolution. Furthermore, those who followed him, like Geoffroy and Chambers, differed widely from Lamarck and from each other in many respects. Lamarck certainly had not effected the replacement of one paradigm by a new one.

By contrast, no one can deny that Darwin's *Origin of Species* (1859) produced a genuine scientific revolution. Indeed it is often called the most important of all scientific revolutions. Yet, it does not at all conform to Kuhn's specifications. The analysis of the Darwinian revo-lution encounters considerable difficulties because his paradigm actu-ally consisted of a whole package of theories, five of which are most important (see Chapter 9).[7] Matters become much clearer if one speaks of Darwin's first and second scientific revolutions.

The first one consisted of the acceptance of evolution by common descent. This theory was revolutionary in two respects. First, it replaced the concept of special creation, a supernatural explanation, with that of gradual evolution, a natural, material explanation. And second, it replaced the model of straight-line evolution, adopted by earlier evo-lutionists, with that of branching descent, requiring only a single origin of life. This was at last a persuasive solution for what numerous authors, from Linnaeus on (and earlier), had attempted to find, a "natural" system. It rejected all supernatural explanations. It furthermore deprived man of his unique position and placed him in the animal series. Common descent was remarkably rapidly adopted and formed the most active and perhaps the most successful research program of the immediate post-Darwinian period. The reason why it fitted so well into the research interests of morphology and systematics is that it supplied a theoretical explanation of previously discovered empirical evidence, such as the Linnaean hierarchy and the archetypes of Owen and von Baer. It did not involve any drastic shift of a paradigm. Furthermore, if one were to accept the period from Buffon (1749) to the *Origin* (1859) as a period of normal science, one would have to deprive a number of smaller revolutions, which took place also within this period, of their revolutionary status. This includes the discovery of the great age of the earth, of extinction, of the replacement of the *scala naturae* by morphological types, of biogeographic regions, of the concreteness of species, and so on. All of these were necessary prerequisites for Darwin's theory and could be included as components of the first Darwinian revolution, shifting the beginning of the first Darwinian revolution back to 1749.[8]

The second Darwinian revolution was caused by the theory of natural selection. Although proposed and fully explained in 1859, it encountered such solid opposition owing to its conflict with five prevailing ideologies (creationism, essentialism, teleology, physicalism, and reductionism) that it was not generally accepted until the evolutionary synthesis of the 1930s and 40s. And in France, Germany, and some other countries there is considerable resistance to it even at the present time.

When did this second Darwinian revolution take place—when it was proposed in 1859 or when it was broadly adopted in the 1940s? Can one consider the period from 1859 to the 1940s a period of normal science? Actually a considerable number of smaller revolutions in biology took place in this period, such as the refutation of an inheritance of acquired characters (Weismann 1883), the rejection of blending inheritance (Mendel 1866, and many subsequent papers), the development of the biological species concept (Poulton, Jordan, Mayr), the discovery of the source of genetic variation (mutation, genetic recombination, diploidy), the appreciation of the importance of stochastic processes in evolution (Gulick, Wright), the founder principle, the proposal of numerous genetic processes of evolutionary consequence, and so on. Many of these had indeed a rather revolutionary impact on the thinking of evolutionists, but without any of the Kuhnian attributes of a scientific revolution.

After the general adoption of the synthetic theory, let us say from 1950 on, modifications of almost all aspects of the paradigm of the synthesis were proposed and some adopted. Nevertheless, there can be little doubt that throughout the period from 1800 to the present there were periods of relative quiet in evolutionary biology, and other periods of rather vigorous change and controversy. In other words, neither the Kuhnian image of well-defined short revolutions and intervening long periods of normal science is correct, nor that of slow, steady, even progress.

It would be interesting, but has not yet been done, to look at breakthroughs in various other fields of biology and see to what extent they qualify as revolutions, whether they led to the replacement of one paradigm by another, and how much time it took before the replacement was completed. For instance, was the origin of ethology (put forth by Lorenz and Tinbergen) or the cell theory (Schwann, Schleiden) a scientific revolution? Perhaps the most revolutionary development of biology in the twentieth century was the rise of molecular biology. It resulted in a new field, with new scientists, new problems, new experimental methods, new journals, new textbooks, and new culture heroes, but conceptually the new field was nothing but a smooth continuation of the developments in genetics preceding 1953; there was no revolution during which the previous science was rejected.[9] There were no incommensurable paradigms. Rather it was the replacement of coarse-grained by fine-grained analysis and the development of entirely new methods. The rise of molecular biology was revolutionary, but it was not a Kuhnian revolution.

GRADUALISM IN BIOLOGICAL ADVANCES

Virtually all authors who have attempted to apply Kuhn's thesis to theory change in biology have found that it is not applicable in this field. Even in the cases in which there was a rather revolutionary change, it did not take place in the form described by Kuhn. First of all, there was no clear-cut difference between revolutions and "normal science." What one finds is a complete gradation between minor and major theory changes. A number of minor revolutions take place even in any of the periods that Kuhn might designate as normal science. Up to a point this is now also admitted by Kuhn, but this realization did not induce him to abandon his distinction between revolutions and normal science.[10]

The introduction of a new paradigm by no means results always in the immediate replacement of the old one. As a result, the new revolutionary theory may exist side by side with the old one. In fact, as many as three or four paradigms may coexist at the same time. For instance, after Darwin and Wallace had proposed natural selection as the mechanism of evolution, saltationism, orthogenesis, and Lamarckism competed with selectionism for the next 80 years.[11] It was not until the evolutionary synthesis of the 1940s that these competing paradigms lost their credibility.

Kuhn makes no distinction between theory changes caused by new discoveries and those that are the result of the development of entirely new concepts. Changes caused by new discoveries usually have much less of an impact on a paradigm than conceptual upheavals. For instance, the ushering in of molecular biology through the discovery of the structure of the double helix had only minor conceptual consequences, and therefore there was virtually no paradigmatic change during the transition from genetics to molecular biology.

The same new theory may be far more revolutionary in some sciences than in others. Plate tectonics supplies a good illustration. That this theory had a revolutionary, one might almost say cataclysmic, effect on geology is obvious. But what about biogeography? As far as avian distributions are concerned, the historical narrative inferred prior to plate tectonics had to be changed hardly at all (a North Atlantic connection in early Tertiary is the only exception) as a result of the adoption of plate tectonics.[12] To be sure, avian distribution in Australonesia did not agree at all with plate tectonic reconstructions, but later geological work showed that the geological reconstructions were faulty, while the revised construction fitted the biological postulates quite well.[13] That there must have been a Pangaea in the Permian-Triassic had been postulated by paleontologists long before the proposal of plate tectonics. In other words, the interpretation of the history of life on earth was not nearly as much affected by the acceptance of plate tectonics as was that of geology.

The major impact of the introduction of a new paradigm may be a massive acceleration of research in the area. This is particularly well illustrated by the explosion of phylogenetic researches after the proposal by Darwin of the theory of common descent. In comparative anatomy as well as in paleontology, much of the research after 1860 was directed to the search for the phylogenetic position of specific taxa, particularly primitive and aberrant ones. There are many other instances where remarkable discoveries had relatively little impact on the theory structure of the field. The unexpected discovery by Meyen and Remak that new cells originate by the division of old cells and not by the conversion of a nucleus into a new cell had remarkably little impact. As far as genetic theory is concerned, likewise, the discovery that the genetic material is nucleic acids rather than proteins did not lead to a major paradigm shift.

The situation is somewhat different with the development of new concepts. When Darwin's theorizing forced the inclusion of man in the tree of common descent, it caused indeed an ideological revolution. By contrast, as was correctly emphasized by Popper (1975), Mendel's new paradigm of inheritance did not. It cannot be overemphasized that changes in concepts have far more impact than new discoveries. For instance, the replacement of essentialistic by population thinking had a revolutionary impact in the fields of systematics, evolutionary biology, and even outside of science (in politics). This shift had a profound effect on the interpretation of gradualism, speciation, macroevolution, natural selection, and racism. The rejection of cosmic teleology and of the authority of the Bible has had equally drastic effects on the interpretation of evolution and adaptation.

Finding virtually no confirmation of Kuhn's thesis in a study of theory changes in biology inevitably forces us to ask what induced Kuhn to propose his thesis? Since much of explanation in physics deals with the effects of universal laws, such as we do not have in biology, it is indeed possible that explanations involving universal laws are subject to Kuhnian revolutions. But we must also remember that Kuhn was a physicist and that his thesis, at least as presented in his early writings, reflects the essentialistic-saltationistic thinking so widespread among physicists. Each paradigm was at that time, for Kuhn, of the nature of a Platonic *eidos* or essence and could change only through its replacement by a new *eidos*. Gradual evolution would be unthinkable in this conceptual framework. Variations of an *eidos* are only "accidents," as it was called by the scholastic philosophers, and therefore variation in the period between paradigm shifts is essentially irrelevant, merely representing normal science.

Does Science Advance through a Darwinian Process?

The picture of theory change that Kuhn painted in 1962 was congenial to the essentialistic thinking of physicalists, but it was incompatible with the thinking of a Darwinian. It is therefore not surprising that the Darwinians favored an entirely different conceptualization for theory change in biology, usually referred to as Darwinian evolutionary epistemology. As Feyerabend (1970) points out, this is actually a very old philosophical concept: "The idea that knowledge can be advanced by a struggle of alternative views and that it depends on proliferation was first put forth by the Presocratics (this has been emphasized by Popper himself) and it was developed into a general philosophy by Mill (especially in *On Liberty*). The idea that a struggle of alternatives is decisive for science, too, was introduced by Mach *(Erkenntnis und Irrtum)* and Boltzmann *(Populärwissenschaftliche Vorlesungem)*, mainly under the impact of Darwinism."

The principal thesis of Darwinian evolutionary epistemology is that science advances very much as does the organic world—through the Darwinian process. Epistemological progress thus is characterized by variation and selection. More precisely, "More robust ideas or ideas with greater verisimilitude or greater explanatory power or greater problem solving ability, etc., survive better from one generation to the next in the struggle for acceptance" (Thompson 1988:235). One can document this process, for instance, in Darwin's own theorizing. In his younger years he proposed one evolutionary theory after the other, always eventually rejecting them until he finally arrived at his evolution by descent through natural selection.[14] Or in the post-Darwinian period one could describe the great variation of evolutionary theories that competed with natural selection—Lamarckism, saltationism, orthogenesis—leaving only natural selection as the successful survivor. There is indeed great similarity with natural selection in the competition among conjectures and hypotheses dealing with an epistemological problem, with one or the other eventually being victorious, at least temporarily. At a superficial level there is no doubt that the historical advance of scientific theories has a strong resemblance to the Darwinian process of evolutionary change.

When analyzed more closely, however, epistemological change actually occurs in a manner that in many respects is different from genuine evolutionary change.[15] The variation, for instance, among the various theories is not caused by chance, as is genetic variation, but by the reasoning of the promoters of these theories. While true, this is not a weighty argument because the source of variation is of little consequence for the Darwinian process. Darwin, for instance, accepted some since-refuted so-called Lamarckian processes, such as "use and disuse" and a direct effect of the environment as sources of new variation. Even in the synthetic theory of the 1940s, many sources of variation are accepted: mutation, recombination, biased variation, horizontal transfer, hybridization, and others. Hence, it is irrelevant whether the variation is produced by chance or not.

The transmission from generation to generation in evolutionary epistemology is cultural transmission, something very different from genetic transmission, to mention only another of a series of differences. Also, the major theory advances ("Kuhnian revolutions") are perhaps more drastic than the genetic changes compatible with the nature of biological populations.

Even though it is thus quite obvious that epistemological changes are not isomorphic with Darwinian evolutionary changes, it is nevertheless true that they do occur according to the basic Darwinian model of variation and selection. In a group of competing theories, the one that will ultimately prevail is the one which runs into the fewest

difficulties and is able to explain the greatest number of facts satisfactorily—in other words, the one that is the "fittest." This is a
Darwinian process. In epistemology, as in biological populations, there
is a continuous production of new variation, that is, of new conjectures. Some of them fit the situation better than others—that is, they
are more successful and will be accepted until modified or replaced
by still better explanations. There is great variation in the size of the
changes—many of them being very minor, others being sufficiently
drastic to deserve to be called revolutions. Branching descent, natural
selection, nucleic acid instead of proteins as carriers of genetic information are among the advances that have had the most revolutionary
impact.

From these observations one can draw the following conclusions:
(1) There are indeed major and minor revolutions in the history of
biology. Yet even the major revolutions do not necessarily represent
sudden, drastic paradigm shifts. (2) An earlier and a subsequent
paradigm may coexist for long periods. They are not necessarily incommensurable. (3) Active branches of biology seem to experience no
periods of "normal science." There is always a series of minor revolutions between the major revolutions. Periods without such revolutions are found only in inactive branches of biology, but it would seem
inappropriate to call such quiet periods "normal science." (4) Darwinian evolutionary epistemology seems to fit theory change in biology
far better than Kuhn's description of scientific revolutions. Active areas
of biology experience a steady proposal of new conjectures (Darwinian
variation) and some of them are more successful than others. One
can say that these are "selected," until replaced by still better ones, or
one can say that inferior or invalid conjectures and theories are
eliminated so that in the end the only theory left is that which is most
successful in explanation. And (5) a prevailing paradigm is likely to
be more strongly affected by a new concept than by a new discovery.

WHY SCIENTIFIC CONSENSUS IS SO HARD TO ACHIEVE

Nonscientists often naively assume that once a new scientific explanation or theory has been proposed, it will quickly be adopted. Actually, cases where a sudden new insight led to a revolutionary instan-

taneous illumination of a field have occurred only very rarely. Most
major tenets of modern science had to overcome years of resistance
both from within and from outside of science. As we have seen,
Darwin's and Wallace's theory of natural selection was not adopted by
the majority of scientists from 1859 until about 1940. Continental
drift was first advanced by Wegener in 1912, although there had already
been a number of forerunners. The geophysicists opposed this theory
almost unanimously, arguing simply that no force was known that
could move entire continents all over the map nor could it explain
the geology of the ocean floor. Some of the biogeographic cases cited
in support of drift (Pleistocene distribution patterns) were badly chosen and easily refuted. However, eventually more and more evidence
for continental drift accumulated, particularly through the researches
of paleontologists, so that when in the early 1960s sea floor spreading
and correlated magnetic phenomena were discovered, continental drift
was accepted within a few years.[16]

Another theory that was proposed long before it was accepted was
geographic speciation (multiplication of species). On the basis of the
Galapagos evidence, Darwin at first (in the 1840s) supported strict
geographic speciation. But later (in the 1850s) he accepted also sympatric speciation, and indeed eventually thought that it was the more
frequent and more important process.[17] Moritz Wagner's view (1864,
1889) that speciation is usually geographic was a minority view until
1942.[18] In the 80 years after 1859, the mapping of the distribution of
subspecies, incipient species, and closely related species of birds, mammals, butterflies, and snails led to the almost universal conviction that
geographic speciation is the major, perhaps almost exclusive mode of
speciation in sexually reproducing organisms. Since that time, so many
new arguments have been advanced in favor of sympatric and other
forms of nongeographical speciation that the question as to whether
these other modes of speciation occur and if so, to what extent, is still
controversial. Conceptual positions are clearly involved in this argument, some authors approaching the problem from the point of
populational geography, while others derive their arguments from local
ecology.

The reasons why some theories have to struggle for the better part of a century before they are accepted, while a few new ideas succeed almost instantaneously, are manifold; I will list six of them.[19]

One reason that consensus takes a long time to achieve is that different sets of evidence lead to different conclusions. For instance, the student of geographic speciation is consistently impressed by the gradualness of the speciation process and considers this powerful evidence for gradual evolution. By contrast, many paleontologists have been equally impressed by the universality in the fossil record of gaps between species as well as between higher taxa and have considered this equally convincing evidence for saltational evolution. The resulting challenge, then, is to show how the discontinuous fossil record can be reconciled with the gradual process of speciation. This was attempted by Mayr, Eldredge and Gould, and Stanley.[20]

A second reason why consensus is hard to achieve is that disagreeing scientists adhere to different underlying ideologies, making certain theories acceptable to one group which are impossible for another group. For instance, the theory of natural selection was unacceptable in 1859 (and ensuing years) to creationists, natural theologians, teleologists, and deterministic physicalists. The replacement of ideologies ("deep paradigms") meets far more resistance than the replacement of erroneous theories. Such viewpoints as vitalism, essentialism, creationism, teleology, and natural theology were an essential part of the worldview of those who held them and were not easily given up. Opposing concepts therefore spread only slowly, by recruiting adherents who did not yet have a firm worldview.

A third reason is that at a given time several explanations may seem to account for the same phenomena equally well. An instance is long-distance orientation in birds, which has been attributed to sun orientation, magnetism, olfaction, and other factors.

In some cases there is actually a pluralism of possible answers. For instance, completed speciation may be achieved by the acquisition of either premating or postmating isolating mechanisms; or relatively rapid geographic speciation may take place either in founder or in relict populations; or species status may be attained by chromosomal reorganization.

Sometimes a consensus cannot be reached because one biologist is concerned with proximate, the other with evolutionary, causations. For T. H. Morgan, sexual dimorphism was explained by the sex chromosomes and hormones (proximate causations), while for the students of evolution it is explained by selection for reproductive success (an evolutionary causation).

Some factors that work against acceptance of new ideas are not strictly scientific. Perhaps one author was disliked or had even offended the current establishment, while another had unexpected success with a subsequently refuted theory because he belonged to a powerful clique. When the scientists involved belong to different schools or countries in which different explanatory schemes have been traditional, consensus may be harder to achieve. Presumably in these cases one of the other five reasons listed above had been primary, but once a tradition was established it was tenaciously maintained even in the face of all opposing evidence. An example is the long-lasting preference by many French authors for a Lamarckian interpretation of evolution, while in most other countries selectionism had already been victorious. The scientific establishment of a country is usually more ready to accept the work of an author of their own nation, or one who has at least published in their language, than the writings of foreign authors. Important work published in Russian, Japanese, or even non-English western European languages is likely to be widely neglected, if not ignored altogether. Even if the ideas contained in such neglected publications are eventually adopted, it is often because someone else rediscovered them subsequently, and the priority of the earlier publication is forgotten.

The Limits of Science

In his famous essay *Ignoramus, ignorabimus* ("We do not know, we will never know") DuBois-Reymond in 1872 listed a number of scientific problems which he was sure science would never be able to solve. Yet by 1887 he had to admit that some had already been solved. Indeed, some of his critics claimed that *all* had been solved in principle or were on the way to solution.

Occasionally one reads the overenthusiastic statement that science can find the solution to all our problems. Every good scientist knows that this is not true.[21] Some of the limitations of science are practical, while others are a matter of principle. There is general agreement that certain experiments with human subjects are out-of-bounds on principle. They violate our moral standards, perhaps even our moral sense. Certain experiments in "big physics," on the other hand, are simply too expensive to justify support. Here again there is a definite limit, though in this case the limitation is one of practicality.

A serious practical limit to science is the difficulty of exhaustively explaining the workings of a highly complex system. I am sure that in due time we will understand, in principle, the workings of development, of the brain, and of an ecosystem. But considering, for instance, the more than 1 billion neurons in the brain, the complete analysis of a particular thought process may forever be too complex for a detailed analysis.

The same practical point can be made about the regulatory mechanisms of the genome, which are highly complex and which are still far from being understood. What is the function (if any) of the vast amounts and different types of noncoding DNA? In some organisms this adds up to more DNA than the total of the coding genes. To assume that all of this DNA is merely an unwanted byproduct ("junk") of various molecular processes is not a palatable solution for a Darwinian. There have been non-Darwinian proposals, but they are not convincing. Here clearly is an area of unfinished science. My guess is that some of the DNA is indeed an unselected (or not yet counter-selected) byproduct of molecular process, but that other components are part of the complex regulatory machinery of the genome.

Most problems relating to "What?" and "How?" questions are, at least in principle, accessible to scientific elucidation. It is different with "Why?" questions. Many of the latter, particularly those relating to the basic properties of molecules, are unanswerable. Why does gold have the color of gold? Why do electromagnetic waves of a certain wavelength produce in our eyes the sensation of redness? Why are rhodopsins the only molecules to have the capacity to translate light into nerve impulses? Why do bodies respond to gravity? Why are atomic nuclei composed of elementary particles?

Some of these are probably solvable by chemistry, quantum mechanics, and molecular biology. But there are other "ultimate questions," particularly pertaining to values, that can never be answered. This includes the many unanswerable questions often asked by nonscientists. "Why do I exist?" "What is the purpose of the world?" and "What was there before the beginning of the universe?" All such questions, and there is an endless number of them, deal with problems outside the domain of science.

The question is sometimes raised as to the future of science. Considering man's unquenchable thirst for knowledge, the incompleteness of our present understanding, and the high success of science-based technology, there is little doubt in my mind that science will continue to flourish and advance as it has for the last 250 years. As Vannevar Bush has said so rightly, science indeed is an endless frontier.

CHAPTER SIX

How Are the Life Sciences
Structured?

Biology, as it exists today, is an extraordinarily diversified science.
Part of the reason is that it deals with exceedingly varied organisms, ranging from viruses and bacteria to fungi, plants, and animals.
It also deals with many hierarchical levels, from organic macromolecules and genes to cells, tissues, organs, and whole organisms, and the
interactions and organization of whole organisms into families, communities, societies, populations, species, and biota. Each level of activity and organization is an area of specialization with its own name—
cytology, anatomy, genetics, systematics, ethology, or ecology, to
mention only a few. Furthermore, biology has a wide range of practical
applications and has given rise to, or is at least involved in, numerous
applied fields such as medicine, public health, agriculture, forestry,
plant and animal breeding, pest control, fisheries, biological oceanography, and so on.

Even though biology as a modern science originated as recently as
the middle of the nineteenth century, its roots, as we have seen, go
back to the ancient Greeks. Two distinct traditions that arose over
2,000 years ago are still recognizable today: the medical tradition,
represented by Hippocrates and his predecessors and followers, and
the natural history tradition. The medical tradition, reaching a climax
in the ancient world in the work of Galen (c. 130–200), led to the
development of anatomy and physiology, while the natural history
tradition, culminating in Aristotle's *History of Animals* and his other

biological works, eventually gave rise to systematics, comparative biology, ecology, and evolutionary biology.

The separation of medicine from natural history continued through the Middle Ages and the Renaissance. The two traditions were, however, linked by botany because this field, although a branch of natural history, focused on the plants believed to have medicinal properties. Indeed, all the leading botanists from the sixteenth to the end of the eighteenth centuries—that is, from Cesalpino to Linnaeus—were physicians, with the single exception of John Ray. The more strictly biological components of medicine, in due time, became anatomy and physiology, and those of natural history became botany and zoology, while paleontology was associated with geology. This classification of the life sciences prevailed from late in the eighteenth until well into the twentieth century.[1]

The Scientific Revolution had only a minor impact on biology. What had the most decisive effect was the discovery in the seventeenth and eighteenth centuries of the almost unimaginable diversity of the faunas and floras in different parts of the world. The rich booty brought back by official voyages and individual explorers (such as the plant-collecting students of Linnaeus) led to the founding of natural history collections and museums and favored an emphasis on systematics (see Chapter 7). Indeed, biology in the age of Linnaeus consisted almost entirely of systematics, except for the study of anatomy and physiology at medical schools.

Almost all work in the life sciences during that period was descriptive. It would be a mistake, however, to consider this period of biology as conceptually sterile. Through the natural history of Buffon, the physiology of Bichat and Magendie, the idealistic morphology of Goethe, the work of Blumenbach and his followers Cuvier, Oken, and Owen, and the speculations of Naturphilosophie, the foundations were laid for most of the subsequent conceptual breakthroughs. Still, in view of the enormous diversity and uniqueness in the living world, a much broader factual basis was needed in biology than in the physical sciences. This was laid not only through systematics but also through comparative anatomy, paleontology, biogeography, and related sciences.

The term biology was introduced into the literature as early as 1800 by Lamarck, Treviranus, and Burdach.[2] But at first there was actually no field of research deserving this name. The term indicated, however, a trend or a goal and signified a turning away from a strictly descriptive, taxonomic preoccupation and a move toward a greater interest in living organisms. Treviranus (1802:4) offers this description: "The subject matter of our investigations will be the various forms and manifestations of life, the conditions and laws controlling their existence, and the causes by which this is effected. The science, which occupies itself with these subjects, we shall designate by the name biology or science of life."

The origins of the science of biology as we know it today took place between 1828 and 1866 and is associated with the names von Baer (embryology), Schwann and Schleiden (cell theory), Müller, Liebig, Helmholtz, DuBois-Reymond, Bernard (physiology), Wallace and Darwin (phylogeny, biogeography, evolutionary theory), and Mendel (genetics). The excitement of this period was capped by the publication of *On the Origin of Species* in 1859. Developments in these 38 years led to most of the subdisciplines of biology that we find today.

The Comparative and Experimental Methods in Biology

From the Greek *kosmos* to modern times, philosophers and scientists have used two major approaches in their search for some underlying order in nature. The first was the search for laws to account for the regularities they observed. The other was the search for "relationship." By this was at first meant not phylogenetic relationship but simply "having items in common." And this could be established only by comparison.

The comparative method achieved its greatest triumph in the work of Cuvier and his associates when they developed comparative morphology. At first this was a purely empirical endeavor, but after the proposal by Darwin in 1859 of the theory of common descent it more and more became a rigorous scientific method. The comparative method turned out to be so successful that it was applied to other biological disciplines, leading to comparative physiology, comparative embryology, comparative psychology, and so on. Modern macrotax-onomy is almost exclusively comparative.

A major impetus to the new science of biology was the invention and development of new instrumentation. Instruments invented by Johannes Müller and his students and by Claude Bernard were decisive in the pioneering developments of physiology. No other instrument, however, had a greater impact on the rise of biology than the steadily improved microscope. This resulted in the development of two new biological disciplines, embryology and cytology.[3]

After 1870 a split developed in biology, the reasons for which were not understood at the time. The biology of evolutionary causations (with its almost exclusive emphasis on phylogeny) was based on comparison and on inferences from observations (called speculation by their opponents). The biology of proximate causations, on the other hand (primarily physiology and experimental embryology), stressed experimental approaches. Representatives of these two schools of biology argued vehemently over which of the two was the right one. Today, of course, it is clear that both sets of questions must be answered.

When it was discovered that the structure and function of cells was the same in animals and plants, and that this was also true for the mode of inheritance of individual characteristics, the old division into botany and zoology no longer made very much sense. This was especially true after the great similarity, indeed virtually identity, of all molecular processes in the two kingdoms was discovered, and after the distinctness of the fungi and prokaryotes from either the animal or plant kingdom had been established. It became increasingly obvious that in a classification of biological concepts one would have to look for new ordering principles, not based on the type of organisms.

After the development of cellular and molecular biology, some people argued that there was now no longer any need for zoology and botany at all. However, in certain areas, such as taxonomy and morphology, there remained a need to deal with animals and plants separately. Development and physiology are, likewise, on the whole rather different in plants and animals, and behavior concerns only animals. No matter how brilliant the advances in molecular biology may be, there continues to exist a vital need for a biology of whole organisms, even though such a biology might have to be organized very differently from the traditional one.

But aside from these exceptions, all biological problems concern plants and animals equally. What is particularly interesting about the origin of the various new biological disciplines is that equivalent contributions were made by students both of plants and of animals. The botanist Brown discovered the cell nucleus, and the botanist Schleiden with the zoologist Schwann proposed the cell theory, further developed by Virchow, who came from zoology and medicine. The problem of fertilization likewise was solved by a series of discoveries made by botanists and zoologists, and this is equally true for cytology and later for genetics.

Numerous attempts have been made to develop a rational classification of all biological disciplines, to deal with the enormous range of phenomena brought together under the heading biology, but none of them has been entirely successful so far. Among all the classifications of biology that have been proposed over time, none has been more misleading than the one that recognized three branches of biology: descriptive, functional, and experimental. Not only were entire fields of biology (like much of evolutionary biology) virtually excluded by this classification, but it ignored the fact that description is a necessity in all parts of biology, and that the experiment is a major tool of analysis almost exclusively in functional biology. Furthermore, the experiment is most important not so much as a means of data collecting but rather for the testing of conjectures.

Driesch revealed how little he understood the structure of biology when he remarked how fortunate it was that at German universities chairs were now given only in experimental biology and none in taxonomy. Here he lumped evolutionary biology, ethology, and ecology with taxonomy, and considered all parts of organismic biology purely descriptive sciences because they were not experimental. Gillispie's comment that taxonomy does not interest the historian is another example of a misconception of different biological disciplines.

New Attempts to Structure Biology

In 1955 the Biology Council organized a special symposium devoted to the analysis of the concepts of biology and how best to represent the structure of biology.[4] The criteria by which various authors proposed to divide biology into disciplines were exceedingly varied. Widely favored was Mainx's division into morphology, physiology, embryology, and a few other standard subjects, often hierarchically subdivided into cytology, histology, whole organ physiology, and so on, on the basis of morphological considerations. Another widely accepted classification, proposed by P. Weiss, chose a more or less hierarchical approach: molecular biology, cellular biology, genetic biology, developmental biology, regulatory biology, group and environmental biology.[5] Many of the review panels of the National Science Foundation were labeled according to this classification. It is interesting (and no surprise) that the experimentalist Weiss lumped all aspects of organismic biology (systematic, evolutionary, environmental, and behavioral biology) under one category, "group and environmental biology," while reserving five categories of equal weight for hierarchical levels below whole organisms.

Generally, the criteria of classification that any given author suggests are greatly influenced by his educational background. If he comes from the physical sciences or was strongly influenced by them, he is likely to stress experiment, reduction, and unitary components and to concentrate on functional processes.[6] By contrast, those biologists who were raised as naturalists tend to stress diversity, uniqueness, populations, systems, inferences from observation, and evolutionary aspects.

In 1970 the Committee on the Life Sciences of the National Academy recognized twelve categories, the last three of which are applied fields: (1) molecular biology and biochemistry, (2) genetics, (3) cell biology, (4) physiology, (5) developmental biology, (6) morphology, (7) evolution and systematic biology, (8) ecology, (9) behavioral biology, (10) nutrition, (11) disease mechanisms, and (12) pharmacology.[7] While this improved on some of the other systems, it too had problems, such as considering systematics and evolutionary biology to be a single discipline.

Eventually, it was realized that the types of questions one asks in scientific research might help in leading to a more logical classification of the biological disciplines. The three big questions are: "What?" "How?" and "Why?"

"WHAT?" QUESTIONS

One cannot do science, any science, without first establishing a solid factual basis—that is, recording the observations and findings on which theories are based. Description thus is a very important aspect of any scientific discipline.

Curiously, attaching the word "descriptive" to a scientific discipline always has had a somewhat pejorative implication. The physiologists tended to call the morphologists' work descriptive, even though, strictly speaking, most of the physiologists' own work was as descriptive as that of the morphologists. Some molecular biologists have confessed embarrassment that so much of the work published in their field is nothing but a recording of facts (descriptive). There is no need for such embarrassment, because molecular biology, being a new field, needs, like all other branches of science, to go through this descriptive phase.

It would be misleading to recognize a separate discipline, descriptive biology. Description is the first step in any branch of biology. Taxonomy, the recognition of species and higher taxa, is no more descriptive than much of molecular or cellular biology, or, for that matter, the genome project. Description should never be maligned, because it is the indispensable foundation of all explanatory and interpretive research in biology.[8]

What is rather surprising is that the taxonomists themselves, prior to Rensch, Mayr, Simpson, and Hennig, had little appreciation of the worth of their own discipline. In a discussion called "Present Tendencies in Biological Theory" the distinguished ant taxonomist W. M. Wheeler (1929:192) said that taxonomy "is the one biological science that has no theory, being merely diagnostics and classification." How wrong this idea was, was made clear, for instance, by the publications of Hennig, Simpson, Ghiselin, Mayr, Bock, Ashlock, and Hull.[9]

All sciences deal both with phenomena and with processes, but in some sciences the study of phenomena prevails, in others the study of processes. The physiologists, concerned with the explanation of the machinery of life, deal almost exclusively with processes. The evolutionary biologists, however, deal also with processes, those that lead to evolutionary changes, particularly to new adaptations and new taxa. But one of the principal concerns of the naturalists has always been the study of the diversity of life. The study of organic diversity is the special concern of many biological disciplines, particularly taxonomy and ecology. It involves an interaction of complex systems, and requires a rather different strategy, for example, from the analysis of simple physiological processes, as studied in the laboratory.

The study of diversity invariably demands precise and comprehensive description as a first step. This is particularly true for taxonomy (including paleontology and parasitology), biogeography, autecology, and all branches of comparative biology (including comparative biochemistry). This descriptive basis permits the comparisons that lead to the generalizations characterizing the various subdisciplines of evolutionary biology. Criticism is justified only when scientists never go beyond description. The most important results of science are the generalizations and theories that are derived from the raw factual material.

In any field, the data-collecting phase is rarely ever completed. Not only does science as a whole have an endless frontier, but so does each of its many subdivisions. Whenever new methods for data collecting become available, whole new horizons open up to view. Examples of this are the advent of electron microscopy in cytology, scuba gear for shallow water research, or new methods for collecting the fauna in the canopy of tropical forests. Invertebrate zoology made major advances when technologies were developed to collect the meiofauna of the bottom layer of the ocean, the pelagic as well as benthic deep sea fauna, and the organisms associated with volcanic hot vents in the ocean deeps.

Looking back at the history of biology, a biologist is almost embarrassed at how neglected were all organisms that were not higher animals or higher plants. For instance, everything that was not clearly an animal was traditionally considered to belong to the domain of botany. Only very recently have biologists realized how different fungi are from plants (indeed, they are more closely related to animals) and even more recently how strikingly different the prokaryotes (bacteria and relatives) are from the eukaryotes (including the protists, fungi, plants, and animals). The Prokaryota are now recognized as a separate super kingdom, and provide a remarkable example of the endless frontier that exists in biology even at the descriptive level.

"HOW?" AND "WHY?" QUESTIONS

Answers to the "What?" questions alone failed to produce a satisfactory solution to the problem of how to classify the subdivisions of biology. Hence, we must now turn to the "How?" and "Why?" questions.[10] In functional biology, as in all aspects of physiology from the molecular level to the function of whole organs, research deals primarily with "How?" questions. How does a particular molecule perform its function? By what pathway does a whole organ function? Such questions, which deal with the here and now, have been referred to as the study of proximate causations. This field, from the molecular level up to whole organisms, deals primarily with the analysis of processes.

"How?" is the most frequent question in the physical sciences, and it led to the discovery of the great natural laws. It was the dominant question also in biology until the early 1800s because the then-leading biological disciplines, physiology and embryology, were dominated by physicalist thinking. These two disciplines were almost exclusively concerned with the study of proximate causations. To be sure, "Why?" questions were also asked, but with Christianity being at that time the dominant ideology of the Western world, such questions inevitably yielded the facile answer: God the Creator (creationism), God the Law-Maker (physicalism), and God the Designer (natural theology).

"Why?" questions deal with the historical and evolutionary factors that account for all aspects of living organisms that exist now or have existed in the past. Why are hummingbirds restricted to the New World? Why are desert animals usually colored like the substrate? Why do insect-eating temperate zone birds migrate in the fall to subtropical or tropical areas? Such questions, usually relating to adaptations or to organic diversity, have traditionally been referred to as the search for ultimate causations. "Why?" questions did not become scientific questions until after the proposal of evolution and more particularly until after 1859 when Darwin proposed a concrete mechanism for change: natural selection.

Very few people realize that it was Darwin who was responsible for making "Why?" questions scientifically legitimate. And by asking these questions he brought all of natural history into science. Physicalists like Herschel and Rutherford had excluded natural history from science because it did not conform to the methodological principles of physics. The nature of inanimate objects, not having a historically acquired genetic program, cannot be elucidated by "Why?" questions. What Darwin did was to add a most important new methodology to the equipment of science.

The terminology of proximate and ultimate causations has a long history, perhaps going back to the days of natural theology, when "ultimate" referred to the hand of God. It has been said that Herbert Spencer spoke of ultimate and proximate causes, but the earliest reference I have been able to find is in a letter which G. J. Romanes (1897:98) wrote to Darwin in 1880: "To offer . . . molecular movements . . . as a full explanation of heredity seems to me like saying that the cause, say, of an obscure disease like diabetes, is the persistence of force. No doubt this is the ultimate cause, but the pathologist requires some more proximate cause if his science is to be of any value."

Considering the vagueness of this statement, it is not surprising that it took another 40 years until a better defined usage was introduced into the literature by John Baker (1938:162). It is of interest to quote in full his use of these terms: "Animals have evolved the capacity to respond to certain stimuli by breeding. In cold and temperate climates it is usually clear that the season adopted allows the young to grow up in favorable climatic conditions, and one may say that in a sense these conditions are the ultimate cause of the breeding season being at that particular time. There is, of course, no reason to suppose that the particular environmental conditions favorable to the young are necessarily the one or ones which constitute the proximate cause and stimulate the parents to reproduce. Thus abundance of insect food for the young might be the ultimate, and length of day the proximate cause of a breeding season."

David Lack (in 1954) took this terminology over from Baker, and I (in 1961) adopted it from both of these authors (even though after Darwin ultimate causation simply meant evolutionary causation). The concept was quickly further developed by Orians (1962) and some ethologists. Even before 1961, perceptive biologists understood well that there are these two sides to biology. Weiss (1947:524), for instance, stated: "All biological systems have a dual aspect. They are causal mechanisms as well as products of evolution . . . Physiology may want to stay on the side of the repeatable and controllable phenomena and leave the singular and non-repetitive cause of historic evolution to others." But neither Weiss nor anyone else enlarged upon these hints until I formalized the distinction in 1961.

Proximate causes relate to the function of an organism and its parts as well as its development, as investigated from functional morphology to biochemistry. They deal with the decoding of genetic and somatic programs. Evolutionary (historical or ultimate) causes, on the other hand, attempt to explain why an organism is the way it is, as a product of evolution. They explain the origin and the history of genetic programs. Proximate causes are usually the answer to the question "How?" while evolutionary causes are usually the answer to the question "Why?"

Unfortunately, through much of the history of biology of the last 130 years endeavors were made to explain biological phenomena exclusively in terms of either one or the other of these two causations. The experimentalists would say that development was entirely due to physiological processes in the developing embryo, while evolutionary biologists would stress that the egg of a fish would always develop into a fish, and that of a frog into a frog, also that such phenomena as recapitulation would not make sense unless the evolutionary aspects were considered. Many of the great controversies in biology of the past, such as the controversy between the nature and nurture schools in inheritance and behavior, or the rebellion of the Entwick-lungsmechaniker against the Haeckelian comparative embryologists,[11] were the result of this one-sidedness.

The continuing confounding of questions dealing with proximate and ultimate causations is particularly apparent in the writings of the so-called structuralists and the idealistic morphologists. Their basic rationale is antiselectionist and rather teleological; they see logic, order, and rationality in the biological realm.[12] Chance, as an explanatory principle, is frowned upon and is always considered an alternative to selective directional processes rather than a simultaneous process. Consideration of the "historical" (evolutionary) component of biological phenomena is to be avoided if at all possible.[13] That both causations have to be considered in most biological explanations, purely physicochemical ones excepted, is not seen by the structuralists.

The recognition that biological inquiry can be broken down into these two very different questions has helped to resolve various conceptual controversies in biology, and it has led to methodological clarification (what method to use when) and to a clearer demarcation between various biological disciplines. It has also called attention to the historical aspect of ultimate causations and to the physiological mechanisms involved in proximate causations, and it has demonstrated that most biologists are on the whole students either of ultimate or of proximate causations, owing to their choice of field in which to work. Yet, as I have always insisted, no biological phenomenon is fully explained until both proximate and ultimate causations are illuminated. Even though most biological disciplines concentrate on either one or the other set of questions, each of these disciplines, to a lesser or greater extent, has to consider also the other type of causations.

Let me illustrate this for molecular biology. A given molecule has a functional role in an organism. How it performs this role, how it interacts with other molecules, its role in the energy balance of the cell, and so on—these questions result in a study of proximate causations. But when we ask why the cell contains this molecule, what role it played in the history of life, how it may have changed during evolution, how and why it differs from homologous molecules in other organisms, and similar questions, then we are dealing with ultimate causations. The study of both kinds of causations is equally legitimate and indispensable.

The study of animal behavior is another area that demonstrates the particularly close connection between the two types of causation. Why a particular type of organism displays the behavior components it does is a result of evolution. But to explain the neurophysiology of a particular behavior requires a study of the proximate causations through neurophysiological studies.

Proximate causations impinge on the phenotype, that is, on morphology and behavior; ultimate causations help explain the genotype and its history. Proximate causations are largely mechanical; ultimate causations are probabilistic. Proximate causations occur here and now, at a particular moment, at a particular stage in the life cycle of an individual, during the lifetime of an individual; ultimate causations have been active over long periods, more specifically in the evolutionary past of a species. Proximate causations involve the decoding of an existing genetic or somatic program; ultimate causations are responsible for the origin of new genetic programs and their changes. The determination of proximate causations is usually facilitated by experimentation, of ultimate causations by inference from historical narratives.

A NEW CLASSIFICATION BASED ON "HOW?" AND "WHY?"

What classification of the life sciences might one adopt if one were to arrange them either with proximate or with evolutionary causations, strictly on the basis of their major concern? All of physiology (organ physiology, cellular physiology, sensory physiology, neurophysiology, endocrinology, and so on), most of molecular biology, functional morphology, developmental biology, and physiological genetics fit best with proximate causations. Evolutionary biology, transmission genetics, ethology, systematics, comparative morphology, and ecology fit best with evolutionary causations.

This tentative division immediately results in certain difficulties, such as the necessity of splitting genetics into transmission (and population) genetics and physiological genetics, or of splitting morphology into functional and comparative morphology. However, these disciplines had already been conceptually separated for a long time, even though covered by a single label. Functional morphology, for instance, is often studied by descriptive morphologists, and students of phylogeny make extensive use of molecular methods. Ecology is hard to place; it deals largely with complex systems, and therefore most ecological problems involve both proximate and ultimate causations. When in the nineteenth century the cell theory was developed by Schwann, Schleiden, and Virchow it was clearly a branch of morphology and so it was still in the heyday of electron microscopy, but modern cellular biology is largely molecular biology.

Power Shifts within Biology

The ongoing restructuring of biology could not take place without a good deal of tension, controversy, and dislocation. Whenever a new subdiscipline became successful, it would fight for its place in the sun and would try to take as much attention and resources as possible away from the established disciplines. Sometimes a new field would establish a virtual monopoly. When I got my Ph.D. in Berlin in 1926, several knowledgeable zoologists advised me to switch to Entwicklungsmechanik, if I were to choose academic zoology as my career. "Spemann fills all the vacant chairs," they told me. DuBois-Reymond never concealed his contempt for the "descriptive zoology" of his teacher Johannes Müller, even though in retrospect the achievements of his own research are by comparison not all that impressive. Whichever field was dominant at any time would try to squeeze out the competing fields and capture as many positions as possible. The last time this happened was when molecular biology had its first flowering. The biochemist George Wald loudly proclaimed that there is only one biology and that it is molecular biology; all of biology is molecular, he said. At several universities in the United States most or all organismic biologists were at that time replaced by molecular biologists. With the physical sciences traditionally favored by Nobel Prizes, in elections to the National Academy, in advisory roles of the government, and by industry, those parts of biology closest to the material and the thinking of the physical sciences were always favored by the government, while other aspects of biology, such as the study of biodiversity, were consistently neglected. The origin of this diversity, one of the two principal problems of evolutionary biology, was almost totally ignored by evolutionary genetics prior to the evolutionary synthesis. Medicine-related biology has, for obvious reasons, always been a favorite among granting agencies. Equivalent projects usually get far higher financing when supported by the National Institutes of Health than when supported by the National Science Foundation.

Botany particularly suffered from these developments. In the days of Linnaeus, botany was the *scientia amabilis,* and right up to the early twentieth century there were many botanists among the leading biologists. This was particularly true for cytology and ecology. All three of Mendel's so-called rediscoverers (DeVries, Correns, and Tschermak) were botanists. But then began a series of setbacks. The study of fungi (mycology) was removed from botany and became an independent field; even more importantly, so did the study of the prokaryotes. Most zoologists, after around 1910, had become specialists in cytology, genetics, neurophysiology, behavior, and so on, and felt that they were dealing with basic biological phenomena and wanted to be called biologists rather than zoologists—a word which, rightly or wrongly, always seemed to remind them of morphology or taxonomy. The word "biological" increasingly often was used comprehensively for the combination of botany and zoology. For example, in 1931 at Harvard the Biological Laboratories were established in a Department of Biology. In this new department there still were professors who taught strictly botanical subjects, like plant morphology, plant physiology, plant taxonomy, and plant reproductive biology, but now they were rubbing elbows with other biologists who specialized in equivalent zoological subjects.

When the American Institute of Biological Science (AIBS) was founded in 1947, it included botany, zoology, and all other biological disciplines. The botanists, however, were apprehensive (with considerable justification) that the unique characteristics of plants would be forgotten if the consolidation into biology went too far. When in 1975 the National Academy reorganized its subdivisions, the section of zoology was abolished and replaced by a section on population biology, evolution, and ecology. The botanists were invited to do likewise but preferred to preserve their section. They maintained that abandoning a section of plant biology would lead to a neglect of the unique properties of plants. A number of botanists, however, left the section of plant biology and joined such general biological sections as the section of genetics or the section of population biology.[14]

But botany has by no means been obliterated. For instance, it has assumed leadership in the study of tropical biology. Herbaria and botanical journals continue to make important contributions to biology, and botany departments are still active at many colleges and universities. Indeed, in the wake of the modern conservation movement, botany is now again more productive than it had been in the preceding period.

Almost invariably the representatives of a new tradition, the founders of a new discipline, think that this makes one of the classical subdivisions of biology obsolete. Actually, even the most traditional branches of biology—systematics, anatomy, embryology, and physiology—are still needed, not just as data banks but also because all of them are endless unfinished frontiers and all of them are still needed to round out our view of the living world. Each discipline seems to have a golden period, and many of them have several. But even after the law of diminishing returns has taken over, there is no justification for abolishing a discipline that has become "classical."[15]

Biology, a Diversified Science

Chapters 1 and 2 emphasized the distinguishing features and concepts of biology as compared with the physical sciences, theology, philosophy, and the humanities. Almost equally important are the conceptual differences within biology. Each branch of biology has its own data bank, its own set of theories, its own conceptual framework, its own textbooks, journals, and scientific societies. To be sure, there are similarities among the biological disciplines that deal with proximate causations, as well as among those that specialize in ultimate causations, but even they differ remarkably in the nature of the prevailing theories and fundamental concepts.

To do such an analysis for all special areas in biology would have required far more space than is available in this volume, and it would have far exceeded my competence. What I will attempt in the chapters that follow, however, is a sample analysis of four fields—systematics, developmental biology, evolution, and ecology—to convey the nature of the struggle among opposing concepts and the relative maturity of the current conceptual framework of these fields.

But before embarking on that task, perhaps I should elaborate on a point made in the Preface—my reasons for not including certain disciplines in my analysis. Some biological disciplines relate to everything that concerns living organisms. This is certainly true for genetics. The genetic program is the underlying factor of everything organisms do. It plays a decisive role in laying down the structure of an organism, its development, its functions, and its activities. Didactically, the most informative way to deal with the concepts of genetics would be to use the history of genetics as the vehicle. This I have tried to do in my *Growth of Biological Thought*. But there I dealt only with transmission genetics. Owing to the rise of molecular biology, the emphasis has now shifted to developmental genetics, and this kind of genetics has virtually become a branch of molecular biology.

More formidable, and perhaps quite insurmountable, are the problems posed by molecular biology. Whether we deal with physiology, development, genetics, neurobiology, or behavior, molecular processes are ultimately responsible for what happens. Some unifying phenomena are already apparent, such as the homeoboxes; others can be dimly perceived. But every time I have attempted to present a bird's-eye view of molecular biology as a whole, I have been overwhelmed by the mass of detail. For this reason no special section of this book is devoted to molecular biology, though in Chapters 8 and 9 I have highlighted some of the major generalizations ("laws") discovered by molecular biologists. The reason that I have not devoted more space to this discipline is not that I consider it less important than other parts of biology—quite the contrary—but that its treatment requires a competence I do not have. The same is true for neurobiology and psychology, which are also exceedingly important. However, I hope that my treatment of biology as a whole will shed some light on those branches of biology that are not covered in detail in this volume.

CHAPTER SEVEN

"What?" Questions:
The Study of Biodiversity

The most impressive aspect of the living world is its diversity. No two individuals in sexually reproducing populations are the same, nor are any two populations, species, or higher taxa. Wherever one looks in nature, one finds uniqueness.

Our knowledge of the diversity of life has been growing exponentially during the last 300 years. It began with the voyages of exploration and the work of individual explorers, whose recorded observations and collections revealed differences in the faunas and floras of every new continent and island explored. Next came the study of freshwater and ocean organisms, including those in the deep sea, which revealed another dimension of biodiversity. The investigation of microscopic plants and animals, parasites, and fossil remains has acquainted us still further with uniqueness in the earth's biota. Finally came the discovery and scientific study of the prokaryotes (bacteria and their relatives), both living and fossil. The particular field of research whose task it became to describe and classify this vast diversity of nature is called taxonomy.

After an initial burst of interest in classification by Aristotle and Theophrastus around 330 BC, taxonomy experienced a long decline until the Renaissance. The field had a second great flowering through the work of Linnaeus (1707–1778), followed by another decline that was halted only when Darwin published his *Origin of Species* in 1859.[1] This work was essentially the result of taxonomic research, and tax-

onomy has continued to play an important role in the development
of evolutionary theory, providing the basis for the biological species
concept and for major theories of speciation and of macroevolution
(see below).

Realizing that the task of studying biodiversity is greater than mere
description and inventory-taking, Simpson suggested that the term
"taxonomy" should be restricted to the traditional aspects of classifying, while the term "systematics" should be applied to "the scientific
study of the kinds and diversity of organisms, and of any and all
relationships among them." Systematics thus was conceived as the
science of diversity, and this new broadened concept has been widely
adopted by biologists.[2]

Systematics includes not only identification and classification of
organisms but also the comparative study of all characteristics of
species as well as an interpretation of the role of lower and higher
taxa in the economy of nature and in evolutionary history. Many
branches of biology depend entirely on systematics; this includes
biogeography, cytogenetics, biological oceanography, stratigraphy, and
certain areas in molecular biology.[3] It is a synthesis of many kinds of
knowledge, theory, and method applied to all aspects of classification.
The ultimate task of the systematist is not merely to describe the
diversity of the living world but also to contribute to its understanding.[4]

Classification in Biology

In daily life, one can deal with a large number of very different items
only by classifying them. Classifications are used for the ordering of
tools, drugs, and art objects, as well as for theories, concepts, and
ideas. When we classify, we group objects into classes according to
their shared attributes. A class, then, is an assemblage of entities that
are similar and related to one another.

Every classification system has two major functions: to facilitate
information retrieval and to serve as the basis of comparative research.
Classification is the key to the system of information storage in any
field. In biology, this information storage system consists of museum

collections and the vast scientific literature in books, journals, and other publications. The quality of any classification scheme is judged by its ability to facilitate the storing of information in relatively homogeneous divisions and to permit rapid discovery and retrieval of this information. Classifications are heuristic systems.

Considering that classifying has been a human activity beginning with our most primitive human ancestors, it is surprising how much uncertainty and disagreement still exists about the nature of classification. And considering how important the process of classifying is in all areas of science, it is curious the extent to which philosophers of science after Whewell (1840) have neglected this subject. However, the person who attempts to classify organisms can derive some elementary rules from everyday human activities such as classifying books in a library or goods in a store: (1) Items that are to be classified should be assembled into classes that are as homogeneous as possible. (2) An individual item is included in the class with whose members it shares the greatest number of attributes. (3) A separate class is established for any item that is too different to be included in one of the previously established classes. (4) The degree of difference among the classes is expressed by arranging them in a hierarchy of nested sets. Each categorical level in the hierarchy represents a certain level of distinctness. These rules apply also to the classification of organisms, even though for the living world some additional rules are required.

Considering how indispensable taxonomic research is to many, if not all, branches of biology, one is surprised by its neglect and low prestige in recent years. The major method in many biological disciplines is comparison, and yet no comparison will lead to meaningful conclusions that is not based on sound taxonomy. In fact, there is no branch of comparative biology—from comparative anatomy and comparative physiology to comparative psychology—that is not ultimately entirely based on the findings of taxonomy.

The multiple roles of taxonomy in biology can be summarized as follows: (1) It is the only science that provides a picture of the existing organic diversity on earth. (2) It provides most of the information needed for a reconstruction of the phylogeny of life. (3) It reveals numerous interesting evolutionary phenomena and makes them available for causal study by other branches of biology. (4) It supplies almost exclusively the information needed for entire branches of biology (such as biogeography and stratigraphy). (5) It supplies ordering systems or classifications that are of great heuristic and explanatory value in most branches of biology, such as evolutionary biochemistry, immunology, ecology, genetics, ethology, and historical geology. (6) Through its foremost exponents, systematics has made important conceptual contributions, such as population thinking (see Chapter 8), that would not otherwise be easily accessible to experimental biologists. These conceptual contributions have significantly broadened biology and have led to a better balance within biological science as a whole.

The taxonomist brings order into the bewildering diversity of nature in two steps. The first is the discrimination of the species, an endeavor referred to as *microtaxonomy*. The second is the classification of these species into related groups, an activity referred to as *macrotaxonomy*. Consequently taxonomy, the combination of the two, was defined by Simpson (1961) as "the theory and practice of delimiting kinds of organisms and of classifying them."

Microtaxonomy: The Demarcation of Species

The recognition, description, and delimitation of species is an activity quite different from other concerns of the taxonomist. It is an area replete with semantic and conceptual difficulties, usually referred to as the "species problem." The term "species" simply means "kind of organism," but because variation is so rampant in the living world, one must define precisely what one means by "kind." A male and a female are also different kinds of organisms, as are infants and adults. As long as it was believed that each species had been separately created, a species was believed to consist of the descendants of the first pair of its kind created by God.

The naturalist dealing with higher organisms such as birds and mammals rarely had any doubt as to what species were. A species for him was simply a group of organisms different from other such groups, where "different" meant differing in visible morphological features.

This species concept was quite widely, indeed almost universally, adopted until the last third of the nineteenth century. Organisms that differed somewhat less than full species were called varieties by Linnaeus and even by Darwin. This species concept was referred to as the typological or essentialistic species concept (and incorrectly as the morphological species concept).

The typological species concept postulated four species characteristics: (1) species consist of similar individuals sharing in the same "essence"; (2) each species is separated from all others by a sharp discontinuity; (3) each species is constant through space and time; and (4) the possible variation within any one species is severely limited. Philosophers referred to such essentialistically conceived species as "natural kinds."

In the course of the nineteenth century the weaknesses of this typological or essentialistic species concept became more and more apparent. Darwin conclusively refuted the notion that species are constant. The studies of geographic variation and particularly the analysis of local population samples confirmed that species are composed of populations which vary from location to location and whose individuals vary within a given population. Types or essences do not exist in living nature.

In addition to these conceptual objections to the typological species concept, there was the purely practical one that it was often of no help in the delimitation of species taxa. Morphological variation within breeding populations, and from one population to another within the same "kind," was often greater than the differences between morphologically similar populations that did not interbreed. Hence a purely morphological criterion was not a reliable criterion for species delimitation. Making matters even worse was the discovery of sibling species—natural populations that were reproductively isolated (that is, not capable of interbreeding because of physiological or behavioral barriers) but which could not be distinguished from one another morphologically. Such populations have now been found in almost all higher taxa of animals and occur also among plants. It became necessary to search for a different criterion for delimiting species, and this was discovered in the reproductive isolation of populations.

From this criterion of noninterbreeding came the so-called biologi-cal species concept. A species, according to this concept, is a group of interbreeding natural populations that is reproductively (genetically) isolated from other such groups because of physiological or behavioral barriers. The only way to fully understand the appropriateness of the biological species concept is to ask Darwinian "Why?" questions: Why are there species? Why do we not find in nature simply an unbroken continuum of similar or more widely diverging individuals, all in principle able to mate with one another? The study of hybrids provides the answer. If the parents are not in the same species (as in the case of horses and asses, for example), their offspring ("mules") will consist of hybrids that are usually more or less sterile and have reduced viability, at least in the second generation. Therefore, there is a selective advantage to any mechanism that will favor the mating of individuals that are closely related (called conspecifics) and prevent mating among more distantly related individuals. This is achieved by the reproductive isolating mechanisms of species. A biological species thus is an insti-tution for the protection of well-balanced, harmonious genotypes.

The biological species concept is called "biological" because it pro-vides a biological reason for the existence of species among organisms, namely, the prevention of interbreeding among incompatible individu-als. It is only incidental that a species may also have other properties, such as the occupation of a separate ecological niche and certain species-specific morphological or behavioral characteristics, that dis-tinguish it from other species.[5]

A major reason for the almost universal acceptance of the biological species concept is its usefulness in most areas of biological research. Ecologists, students of behavior, students of local biota, and even physiologists and molecular biologists are interested in the kinds of populations that can coexist without interbreeding. In many cases, the students of living organisms recognize species not by the morphologi-cal criteria of the typologist but by aspects of their behavior, their life history, or their molecules.

The biological species definition can be applied without difficulty whenever populations in breeding condition coexist at the same lo-cality. It runs into difficulties, however, under two kinds of circumstances. The first is the case of uniparentally reproducing (asexual) organisms, which have no populations and do not interbreed. The biological species concept obviously cannot be applied to such organisms. Exactly what the best criteria are for discriminating species among asexual organisms is not yet clear. Degrees of morphological difference between clones as well as differences in niche utilization have been suggested but have not been adequately tested. Such agamic species are placed in the species category in the Linnaean hierarchy.

The second problem with applying the biological species concept to the delimitation of species is that populations within a species are rarely confined to one limited geographic locality. Rather, they usually extend over a lesser or greater range. When such populations are visibly different from one another, they are usually recognized as subspecies. Subspecies are often part of a continuous series of populations and as such freely interbreed and exchange genes. But many subspecies are geographically isolated and have no opportunity for gene exchange, and as a result they diverge morphologically. Over time such subspecies may eventually attain full species status, because they have acquired a new set of isolating mechanisms. A species consisting of a number of subspecies is called a polytypic species. Species that are not subdivided into subspecies are called monotypic species.

When some of the more distant populations are geographically completely isolated from all other populations within a species, the question arises: Are these isolated populations still members of the parental species? What criteria can one use to decide which of these populations to recognize as full species and which others to combine into a polytypic species? The species status of geographically isolated populations can be determined only by inference, particularly by degree of morphological difference.[6]

The biological species concept was long in coming. Buffon understood the gist of it,[7] and Darwin, in his Transmutation notebooks, said that species status "is simply an instinctive impulse to keep separate." He referred to the "mutual repugnance" of species to intercrossing and pointed out that good species might "differ scarcely in any external character," in other words, that species status had little if anything to do with degree of morphological difference. Curiously, in his later writings, Darwin gave up this biological concept and reverted to a largely typological one.

In the second half of the nineteenth century and the first third of the twentieth, more and more naturalists referred to species in terms of their biological characteristics. Even though they did not propose a formal definition, authors like Poulton, K. Jordan, and Stresemann evidently subscribed to a biological species concept. It was, however, not generally adopted until I proposed a formal definition in 1940 and provided massive support for the biological species concept in my 1942 book, *Systematics and the Origin of Species.*

What helped the acceptance of the biological species concept more than anything else was the vulnerability of the competing concepts. These included the nominalist species concept, the evolutionary species concept, the phylogenetic species concept, and the recognition species concept. None of them is as practical as the biological species concept in delimiting species, though each still has a number of adherents today.

COMPETING SPECIES CONCEPTS

According to the nominalist species concept, only individuals exist in nature, and species are a human artifact; that is, a person (not nature) makes species by grouping individuals under a name. But such arbitrariness is unsubstantiated by the situation encountered in any actual exploration of the natural world. A naturalist who sees, for instance, the four common species of titmice in a British woodland or the common species of wood warblers in a New England forest knows that there is nothing arbitrary about species borders, but that these species are products of nature. Nothing brought this point home to me more forcefully than the fact that the Stone Age primitive natives in the mountains of New Guinea discriminate and name exactly the same species that are distinguished by the naturalists of the West. It requires a vast ignorance of both living organisms and human behavior to adopt the nominalist species concept.

The evolutionary species concept has been promoted particularly by paleontologists who follow species through the time dimension. According to Simpson's (1961:153) definition, "An evolutionary species is a lineage (an ancestral-descendant lineage of populations) evolving separately from others and with its own unitary evolutionary role and tendencies." The main problem with this definition is that it applies equally to almost any isolated population. Also, a lineage is not a population. Furthermore, it side-steps the crucial question of what a "unitary role" is and why phyletic lines do not interbreed with one another. Finally, it actually fails in its objective, the delimitation of species taxa in the time dimension, because in a single gradually evolving phyletic lineage the evolutionary species concept does not permit one to determine at what point a new species begins and where it ends and which part of such a lineage has a "unitary role." The evolutionary species definition ignores the core of the species problem: the causation and maintenance of discontinuities among contemporary living species. It is rather an endeavor to demarcate taxa of fossil species, but it fails even in that endeavor.

The evolutionary species definition ignores the fact that there are two processes by which new species may originate: (1) the gradual change of a phyletic lineage into a different species without changing the number of species, and (2) the multiplication of species through geographic isolation (such as Darwin saw on the Galápagos Islands). The difficulties a taxonomist encounters are almost invariably caused by the latter—the multiplication of species in the horizontal (space) dimension, rather than the change of species in the vertical (time) dimension. The biological species definition specifically addresses the problem of the multiplication of species, while the evolutionary species definition ignores it, dealing only with phyletic evolution. Ordinarily when we speak of speciation, we mean the multiplication of species.

According to the phylogenetic species concept, adopted by many cladists (see below), a new species originates when a new "apomorphy" originates in any population. This apomorphy may be as small as a single gene mutation. Rosen, finding that the species of fishes in almost any tributary of the Central American rivers had locally endemic genes, proposed that all these populations be raised to species rank.[8] One of his critics quite rightly remarked that with the high frequency of neutral gene mutations, every individual is apt to differ from its parents by at least one gene. How would one then decide when a population was different enough to be considered a separate species? This observation clearly showed the absurdity of trying to apply the cladistic concepts of macrotaxonomy to the species problem (for more on cladistics, see below).

The recognition species concept, proposed by H. Paterson, is nothing but a different version of the biological species concept, misunderstood by Paterson.[9]

SPECIES CONCEPT, SPECIES CATEGORY, AND SPECIES TAXA

The word "species" is applied to three very different objects or phenomena: (1) the species concept, (2) the species category, and (3) species taxa. Endless confusion in the literature has resulted from the failure of some authors to discriminate among these three very different meanings of the word "species."

The species concept is the biological meaning or definition of the word "species." The species category is a particular rank in the Linnaean hierarchy—the traditional hierarchy in which organisms are placed. Each rank in this hierarchy (such as species, genus, order, and so on) is referred to as a category. To determine whether a population belongs in the species category, one tests it against the species definition. Species taxa are particular populations or groups of populations that comply with the species definition; they are particulars ("individuals") and thus cannot be defined, only described and demarcated against each other.

At the time of Linnaeus, the identification of species was of concern primarily to the taxonomist, but this is no longer the case. Evolutionary biologists now know that the species is the crucial entity of evolution. Each species is a biological experiment, and there is no way to predict, as far as an incipient species is concerned, whether the new niche it enters is a dead end or the entrance into a large new adaptive zone. Even though evolutionists may speak of broad phenomena such as trends, adaptations, specializations, and regressions, they are not separable from the progression of the entities that display these trends, the species. Owing to their reproductive isolation, whatever evolutionary processes take place in a species are restricted to this species and its descendants. This is why the species is the coin of evolutionary change.

The species is also, to a large extent, the basic unit of ecology. No ecosystem can be fully understood until it has been dissected into its component species, and until the diversified interactions of these species are understood. A species, regardless of the individuals of which it is composed, interacts as a unit with other species with which it shares the environment.

In the case of animals, species are also important units in the behavioral sciences. Members of a species share many species-specific behavior patterns, particularly all those that have to do with social behavior. Individuals that belong to the same species share the same signaling systems in their courtship behavior, and communication systems are largely species-specific. In olfactory species this includes the possession of species-specific pheromones.

The species represents an important level in the hierarchy of biological systems. It is an immensely useful ordering device for many significant biological phenomena. Even though there is no name for the "science of species" (comparable to the name "cytology" for the science of cells), there is no doubt that such a science exists, and that it is one of the most active areas of modern biology.

Macrotaxonomy: The Classification of Species

The branch of taxonomy that deals with the classification (or grouping) of organisms above the species level is called macrotaxonomy. Fortunately, most species seem to fall into natural, easily recognized higher groups, such as mammals and birds or butterflies and beetles. But what is one to do with species that seem intermediate between groups or do not seem to belong to any group?

In the course of the history of taxonomy, there have been many proposals of methods and principles for classifying organisms. The classifications resulting from these principles sometimes had rather different objectives, and this is, perhaps, the reason why even today there is no consensus among taxonomists as to which is the "best" method of classifying.

DOWNWARD CLASSIFICATION

Downward classification was the prevailing method of classification when medicinal botany flourished during and after the Renaissance. Its primary purpose was the identification of different types of plants and animals. At this time knowledge of species in both botany and zoology was still very primitive, and yet it was vitally important to identify correctly the plant which had the known healing properties.

Downward classification proceeds by dividing large classes into subsets, through the use of Aristotle's method of logical division. Animals are either warm-blooded or not; this produces two classes. Warm-blooded animals are either hairy or have feathers, and each of the resulting classes (mammals and birds) can again be subdivided by the process of dichotomy until finally one has arrived at the particular species to which the specimen belongs that one was attempting to identify.

The principles of downward classification dominated taxonomy up to the end of the eighteenth century and is reflected in the keys and classifications proposed by Linnaeus. The method is still used today in field guides and in the keys of taxonomic revisions, except that the method is no longer referred to as classification but as what it really is, which is identification.

Identification schemes had a number of serious weaknesses that prevented them from being useful as true classification systems. They relied entirely on single characters (a "character" in biology is a distinguishing feature or attribute, what we would in everyday usage call a characteristic), and the sequence of characters arbitrarily chosen by the taxonomist controlled the classes produced by the dichotomous divisions. Any gradual improvement of such a classification was nearly impossible, and the choice of certain characters sometimes resulted in highly heterogeneous ("unnatural") groups.

People had, of course, long recognized natural groupings such as fish and reptiles, or ferns, mosses, and conifers. Toward the end of the eighteenth century efforts were made to replace the largely artificial scheme of Linnaeus with a more natural system based on commonly observed similarities and relationships. But there was great uncertainty as to how to determine these criteria.

UPWARD CLASSIFICATION

From about 1770 on, even Linnaeus as well as other taxonomists such as Adanson promoted upward classification as a more appropriate approach. Upward classification consists of assembling species by inspection into groups (taxa) consisting of similar or related species. The most similar of such newly formed taxa are then combined into a higher taxon of the next higher rank until a complete hierarchy of taxa has been formed. This method was simply the application of everyday classification methods to the grouping of species of organisms.

But proponents of upward classification failed to develop a rigorous methodology. There was still a strong tendency to give special weight to conspicuous single characters, and there was no theory to account for the existence of reasonably well defined groups nor for the existence of the hierarchy of taxa. Every taxonomist more or less developed his own methodology.

The years from about 1770 to 1859 was a transition period. The downward method was clearly abandoned, but upward classification was without a well-articulated methodology and often was employed arbitrarily. A subcategory of upward classification developed during this time which consisted of the so-called special-purpose classifications. These classifications were not based on the totality of characters but, for the sake of a special purpose, were based on only one or a restricted number of characters. For instance, mushrooms might be classified for culinary purposes into edible and inedible (or poisonous) ones. Special-purpose classifications go back at least as far as Theophrastus, who distinguished plants according to their growth form into trees, shrubs, herbs, and grasses. Special-purpose classifications are still useful in ecology. For example, a limnologist may divide plankton organisms into autotrophs, herbivores, predators, and detritus feeders. All such systems have a lower information content than a Darwinian classification system.

EVOLUTIONARY OR DARWINIAN CLASSIFICATION

In the brilliant thirteenth chapter of *On the Origin of Species* Darwin put all these taxonomic uncertainties to rest by showing that a sound classification of organisms must be based on two criteria: genealogy (common descent) and degree of similarity (amount of evolutionary change). A classification based on both of these criteria is called an evolutionary or Darwinian classification system.

Philosophers and practical classifiers had long appreciated that if explanatory (causal) theories exist for the grouping of objects, then these explanations must be taken into consideration in the delimitation of such groups. Accordingly, the eighteenth-century classifications of human diseases were replaced in the nineteenth and twentieth centuries with systems based on the etiology of these diseases. Diseases were classified into those caused by infectious agents, by defective genes, by aging, by malignancy, by toxic substances or harmful radiation, and so on. Any classification which takes causation into account is subject to severe constraints that prevent it from becoming a purely artificial system.

As soon as Darwin developed his theory of common descent, he realized that each natural "taxon" (or distinct group of organisms) consists of the descendants of the nearest common ancestor; such a taxon is called monophyletic.[10] If a classification system is based strictly and exclusively on the monophyly of the included taxa, it is a genealogical ordering system.

But Darwin saw very clearly that genealogy "by itself does not give classification." Classifying organisms exclusively on the basis of genealogy is, in a way, merely a special-purpose classification. The criterion of descent was, for Darwin, not a replacement of the criterion of similarity but rather a constraint on the kind of similarity that could be accepted as evidence for relationship. The reason why similarity cannot be neglected is that the diverging branches of the phylogenetic tree "undergo different degrees of modification," and this "is expressed by the forms being ranked under different genera, families, sections or orders" (Darwin 1859:420). In other words, the degree of difference that arises during phylogenetic divergence must be duly considered in the delimitation and ranking of taxa, in order to produce a true classification. A sound Darwinian classification, thus, must be based on a balanced consideration of genealogy and similarity (degree of difference).

To understand the role of similarity in a Darwinian classification, one must understand the concept of homology. Relationship among species and higher taxa is indicated by the existence of homologous characters. A feature in two or more taxa is homologous when it is derived phylogenetically from the same (or a corresponding) feature of their nearest common ancestor. Many kinds of evidence can be used to infer homology. These include the position of the structure in relation to neighboring ones; the connection of two dissimilar stages by an intermediate stage in a related form; similarity in ontogeny; the existence of intermediate conditions in fossil ancestors; and the comparative study of related monophyletic taxa.[11]

But not all similarities between organisms result from homology. Three kinds of character changes during evolution can mimic homology; they are usually grouped together under the term homoplasy. They are convergence, parallelism, and reversal. Convergence is the independent acquisition of the same feature by unrelated evolutionary lineages, such as the acquisition of wings by both birds and bats. Parallelism is the independent realization of a character in two related lineages owing to a genetic predisposition for this character, even though it was not phenotypically expressed in the common ancestor. A well-known example is the independent acquisition of stalked eyes in a group of acalypterate flies. Reversal is the independent loss of the same advanced character in several lineages of a phylogeny. A genealogical analysis would permit the untangling of these similarities among a given group of organisms and the removal from a taxon of those species (or higher taxa) whose similarities are not due to common descent.

The reason why Darwin includes degree of similarity among the classifying criteria is that branching and divergence are not absolutely correlated. There are branching patterns ("trees") in which all branches diverge at about the same rate. Although not exactly true for trees of the language families, they do tend in that direction. The reason is that factors responsible for the evolution of languages are not adaptive but stochastic. When the Anglo-Saxons crossed the North Sea and colonized England, their language did not have to become adapted to the British climate or to political changes. However, when a branch of reptiles (dinosaurs) conquered the air niche, it had to become adapted to the new way of life, and this resulted in a drastic modification of its phenotype. Related branches of dinosaurs who remained in the ancestral niche hardly changed at all. This consideration of ecological factors and their impact on the phenotype characterizes a Darwinian classification.

Until 1965 Darwinian classification was the system in use almost universally, and it continues to be popular today.[12] The delimitation and grouping of related species through similarity is the first step of the process, and the testing of the monophyly of these groups and their genealogical arrangement is the second step. This is the only way by which both of Darwin's criteria for a sound classification of organisms can be satisfied.[13]

A difficulty encountered by the taxonomist is the discordant evolution of different sets of characters. Entirely different classifications, for instance, may result from the use of characters of different stages of the life cycle, such as larval versus adult characters. In the study of a group of bees, Michener (1977) obtained four different classifications when he sorted these species into similarity classes on the basis of the characters of: (1) larvae, (2) pupae, (3) external morphology of adults, and (4) male genitalia. Almost invariably, when a taxonomist makes use of a new set of characters, it leads to a new delimitation of taxa or a change in rank. Even the characteristics of a single stage of the life cycle may change at very unequal rates during evolution.

For instance, when one compares humans with their nearest relatives, the chimpanzees, one finds that *Homo* is more similar in certain molecular characters to *Pan* than are some congeneric species of *Drosophila* to one another. Yet, as we all know, humans differ from even this closest relative among the anthropoid apes very drastically in certain traditional characters (central nervous system and its capacities) and in the occupation of a highly distinct adaptive zone. Almost any organ system and group of molecules in a phyletic lineage will have a somewhat different rate of change from all others. These rates are not constant but may speed up or slow down in the course of evolution. Certain DNA changes are five times as fast in a group of rodents as in the primates, for example. The different rates of evolution of different components of the phenotype require great caution in choosing the characters on which a classification is to be based. The use of different sets of characters may lead to rather different classifications.

Each rank (such as species, genus, order, and so on) in the traditional Linnaean hierarchy is referred to as a category.[14] The lower the rank of a given taxon (group) of organisms, the more similar the included species usually are and the more recent their common ancestor is. There are no operationally defined definitions for any of the higher categories. Many higher taxa are extremely well delimited and can be described unambiguously and with high accuracy (for instance, birds or penguins), but the category in which they are placed is often subjective and involves an element of judgment. A particular group of genera may be called a tribe by some authors, whereas others would call it a subfamily or family.

Most current classifications were developed during the heyday of comparative anatomy in the immediate post-Darwinian period. At that time, when an ancestor was looked for, it was conceived as representing not a single ancestral stem species but a whole taxon. Hence, the nearest common ancestor of the same or lower categorical rank of the mammals is the therapsid reptiles and that of the birds is the dinosaurs (or some other group of reptiles). Owing to this concept and definition of monophyly, all taxa in traditional taxonomy (when correctly formed) were monophyletic. Also, under this concept of monophyly, no group is paraphyletic. For a cladist, a group is paraphyletic if it contains the stem clade (branch) of a derived taxon. The concept of paraphyly makes no sense in a Darwinian classification. For Darwin, then, a taxon was monophyletic if all of its members descended from the nearest common ancestral taxon of the same or lower categorical rank, and this definition is still maintained by Darwinian taxonomists today.

A typical Linnaean hierarchy is characterized by a good deal of discontinuity. Among living organisms there is no intermediacy between reptiles and mammals, nor between tubinares and penguins, nor between turbellarians and trematodes. This observation has long been puzzling and has inspired a number of non-Darwinian saltational theories. Evolutionary researches, however, have helped in providing an understanding of the pattern of diversity.

Most new types of organisms do not originate by the gradual transformation of a phyletic lineage, that is, of an existing type. Rather, a founder species enters a new adaptive zone and succeeds in the new environment by making rapid adaptational adjustments for optimal fitness. Once it has achieved this, the new lineage may enter a period of stasis in which there may be a good deal of speciation but no reconstruction of the structural type (bauplan). The 2,000+ species of Drosophila illustrate this situation. The 5,000+ species of songbirds are also only variations on a single theme.

The two evolutionary processes that produce species— phenotypic change over time and increase in diversity (speciation)—are only loosely correlated. In the traditional Linnaean hierarchy the gaps between taxa and the great variation in size of the higher taxa are explained by this lack of correlation. When a founder species reaches a highly suitable adaptive zone, it may experience copious speciation without experiencing any selection pressure for a change of the basic structural type.[15] The Darwinian classification system is particularly well suited to cope with taxa of highly uneven size and to reflect the gaps between ancestral and derived taxa.

But problems for the Darwinian classifier arise when "horizontal" classification of living taxa is expanded to include the extinct biota. The recent biota consist of the endpoints of countless branches of the evolutionary tree. Higher taxa are separated from one another by gaps caused by divergent evolution and by extinction. Yet, a complete classification of organisms must include extinct groups, all of which are related by descent to one another and to the living biota. The classification of fossil taxa raises numerous problems about which no consensus has yet been reached. How should one treat fossil taxa that are intermediate between two living ones? New taxa, almost invariably, originate by "budding," with the ancestral taxon continuing to flourish. The fossil record is generally far too incomplete to provide evidence for the "stem species" of a derived new taxon.

The two-criteria approach of Darwinian classification—genealogy and similarity—was essentially unchallenged from 1859 until the middle of the twentieth century. To be sure, many taxonomists did not fully practice the conscientious adherence to monophyly testing and a careful weighting of similarity. Entirely new methods, however, were not proposed until the 1960s. Each of the new methodologies makes use of only one of Darwin's two criteria: numerical phenetics is based on similarity, while cladification (Hennigian ordering) is based on genealogy.

NUMERICAL PHENETICS

The objectives of the numerical pheneticists are to avoid all subjectivity and arbitrariness by sorting species, with numerical methods, into groups agreeing in a large number of joint characteristics. Pheneticists believe that the descendants of a common ancestor will share such a multiplicity of characters that they will automatically form well-defined taxa.

Important objections to numerical phenetics are that it is a cumbersome method requiring the analysis of very large numbers of characters (more than 50, preferably more than 100); that it fails to give different weight to characters of different taxonomic importance; that it does not have a methodology for the ranking of taxa; that it fails to allow for different evolutionary rates in different character complexes; that its methods produce different classifications when different character sets are used; and that it cannot be improved gradually.

As long as only morphological characters were available, numerical phenetics was unsatisfactory because there were simply not enough characters to count. When large numbers of molecular characters became available, the situation changed considerably. DNA hybridization is actually a phenetic method, but it avoids most of the standard shortcomings of phenetic analysis owing to the very large number of characters that are taken into account. Some of the "distance" methods of computer taxonomy are also essentially phenetic methods. There is still considerable disagreement as to the value of these methods, compared with other approaches (such as parsimony).

CLADIFICATION

The other recent alternative to Darwinian classification is an ordering system relying entirely on genealogy. In 1950 Willi Hennig published in German a method which, he claimed, would permit the establishment of an unambiguous genealogical classification. His most basic criteria were these: only groups based exclusively on the possession of unquestioned "apomorphies," that is, shared derived characters, should be recognized, while ancestral ("plesiomorphic") characters should be ignored. Furthermore, each taxon should consist of a branch of the phylogenetic tree containing the stem species of this branch and all of its descendants, including all "ex-groups," that is, drastically modified descendants, such as birds and mammals from the reptiles. Hennig's reference system, thus, consists simply of branches (clades) of the phylogenetic tree, without giving any consideration to similarity (that is, amount of evolutionary change).

In the Darwinian evaluation of similarity, as many characters as possible are used, not only apomorphies. Hence, ancestral (plesiomorphic) characters are given appropriate consideration because they often contribute strongly to the aspect, and hence the classificatory status, of a taxon. The same is true for a consideration of autapomorphies in the ranking of sister taxa. The use of as many characters as possible gives a Darwinian classification an additional virtue: "Assigning an object to a particular classification [should] tell us as much as possible about that object. For an extreme perception of order, the ideal would be that correct classification should potentially tell us everything about an object" (Dupré 1993:18).

The Darwinian classification shares with cladistics, in contrast to strict numerical phenetics, the conviction that the cause of the grouping must be given due consideration. Consequently, these two schools of macrotaxonomy insist that the taxa recognized by them must be monophyletic. According to the traditional definition, a taxon is monophyletic if all of its members descended from the nearest common ancestral taxon, and this is the definition still maintained by the Darwinian taxonomists. Hennig, however, proposed an entirely different principle. For him, a group is "monophyletic" when it is composed of all the descendants of the stem species. Since this definition leads to an entirely different delimitation of taxa, Ashlock (1971) proposed the term "holophyletic" for Hennig's new concept. The traditional term monophyletic is a qualifying adjective for a taxon, while Hennig's concept of holophyly refers to a method for delimiting taxa. Even though the taxa delimited by the traditional method may differ from the cladons delimited by Hennig's method, both hierarchies of taxa are strictly genealogical.

A clade of the Hennigian system does not correspond to a taxon of the Darwinian classification and should therefore be given a different technical name, "cladon."[16] Each cladon is traced back to (and includes) the "stem species," that is, the species which displays the first apomorphic character of this branch (clade). Since clades rather than classes form the basis of Hennig's system, it may be distinguished from genuine classifications by the term "cladification."

The methodology of partitioning characters into those that are uniquely derived and those that are ancestral, the so-called cladistic analysis, is an excellent method of phylogenetic analysis. It is a suitable way to test taxa for monophyly. Anyone interested in the phylogenetic aspects of characters will find cladification an excellent method for the ordering of species and taxa with reference to their phylogeny. However, as valuable as a cladogram is for phylogenetic studies, it violates almost all the principles of a traditional classification. Among its deficiencies are the following:

(1) Most clades (cladons) are highly heterogeneous, with the stem species and other stem groups being far more similar to the stem groups of sister clades than to the crown groups of their own clade. In other words, dissimilar groups of species are combined into one cladon, and similar groups of species (sister stem groups) are separated into different cladons.

(2) Either the stem species or the entire stem group very often has traditionally been included in an ancestral taxon, like the therapsid reptiles, the ancestors of mammals, among the Reptilia, and the dinosaurs, the presumed ancestors of the birds, also among the Reptilia. Taking these stem groups out of the taxon with which until now they had always been associated makes this taxon "paraphyletic," and, according to cladistic principles, invalid as a taxon. The result is a

destruction of a high fraction of all the currently recognized higher taxa, and this includes all now recognized fossil taxa that have given rise to derived taxa.

(3) The requirement that sister groups should be assigned the same taxonomic rank is unrealistic because sister groups frequently, if not usually, differ in the number of autapomorphic characters, that is, derived characters, restricted to this particular branch. A sister group that has evolved very little since its origin and one which has undergone a drastic evolutionary transformation (for instance, the birds) had to be given the same categorical rank in the original Hennigian arrangement.

(4) There is no valid theory of ranking in Hennig's methodology. His own followers have abandoned Hennig's only two ranking criteria, geological time and categorical equality of sister groups. Instead, they have adopted the one criterion which Hennig himself specifically rejected, degree of difference but have only subjective criteria for their evaluation.

(5) According to Hennig, every new synapomorphy (derived character) in a stem species requires the assignment of a new categorical rank. Although ignored by the majority of the cladists, some of them have applied this principle at the species level and have gone so far as to demand that every population be raised to species level if it differs even by a single character (the phylogenetic species concept). Such a pulverization of the system would, of course, lead to taxonomic chaos and make any information retrieval virtually impossible.

(6) All nonapomorphic characters are neglected. It is one of the oldest and most often confirmed rules of taxonomy that the more characters one utilized in a classification, the more useful and reliable, on the whole, such a classification will be. Even though quite rightly only derived characters can be used for a cladistic analysis, such a restriction makes no sense when it comes to delimiting taxa in a classification. Indeed, many taxa are characterized by the prevalence of ancestral characters. Furthermore, it completely conceals evolutionary asymmetry in rates of evolution if autapomorphic characters are ignored. It has become evident that a Hennigian cladification actually has the characteristics of an identification scheme rather than those

of a traditional classification. Indeed, leading cladists have again and again stressed that their methodology is a search for characters with diagnostic value.

(7) Cladons, as delimited by a cladist, reflect a one-sided relationship because sister groups, even though genetically more closely related than far-distant descendants, are excluded from the cladons. According to cladistic principles, the modern descendants of Charlemagne are more closely related to him than he was to his brothers and sisters.

In principle, a cladistic classification is a single-character classification. The clade, or "cladon," is characterized by the first apomorphy of the stem species.[17] Any single-character classification, even when strictly complying with phylogeny, results in artificial, heterogeneous taxa. Leading taxonomists for more than a hundred years have rejected single-character classifications. A good classification, they have said, is based on the greatest number of possible characters.

These shortcomings of Hennig's phylogenetic cladification show why it cannot take the place of a traditional Darwinian classification. However, if one is interested only in phylogenetic information, then one should use Hennig's system. In other words, both Hennig's cladification and the traditional Darwinian classification are legitimate but have very different applications and objectives.[18]

Storing and Retrieving Information

In view of all these difficulties, it occurs not infrequently that different authors defend different classifications. Which one should one choose? The answer is, one should choose the most practical one and the one that is most apt to maintain stability in information storage and retrieval. Stability is one of the basic prerequisites of any communication system; the usefulness of a classification stands in direct relation to its stability. The traditional Darwinian system of classification tends to be very stable and is therefore ideal from this point of view. Cladifications, by contrast, are frequently in conflict with traditional classifications, and the study of new characters, as well as a new resolution of homoplasies, may result in considerably modified cladifications, hence in instability.

The sequence of taxa in a collection or printed classification has to be linear (one-dimensional) by necessity, but common descent is a three-dimensional branching phenomenon. It is somewhat arbitrary how one cuts a phylogenetic tree into its branches and twigs and arranges these twigs into a linear sequence. This is particularly true when the phylogenetic tree is a bush (thamnogram) rather than a tree (dendrogram). A number of conventions have been adopted to solve this problem: (1) Place obviously derived taxa after those from which they were derived, hence trematodes and cestodes after the turbellarians. (2) List specialized taxa after the more generalized, seemingly more "primitive" taxa. (3) Avoid changing any widely adopted sequence without cogent reasons, because such a traditional sequence is important for information storage and retrieval, having been adopted in the taxonomic literature and in collections.[19]

NAMES

The names for the higher taxa serve as convenient labels for the purpose of information retrieval, and terms such as Coleoptera and Papilionidae must mean the same thing to zoologists all over the world to have maximum usefulness.[20] It would be quite impossible to refer to the millions of organisms, and to store information about them, if there was not an efficient and universally adopted system of name-giving. For these practical reasons, taxonomists have adopted a number of rules dealing with the provision of names.

These rules are laid down in international codes of zoological, botanical, and microbial nomenclature. The major objectives of the communication system of taxonomists are well stated in the Preamble of the *Code of Zoological Nomenclature* (1985): "The object of the code is to promote stability and universality in the scientific names of animals, and to ensure that each name is unique and distinct. All its provisions are subservient to these ends." The scientific name of a plant or animal is composed of a generic and a specific epithet (Linnaean binomial nomenclature). For instance, the orange hawkweed is *Hieracium* (generic) *aurantiacum* (specific epithet). The language chosen for the scientific names of organisms is Latin, a lingua franca among scientists in the period after the Middle Ages.

Original descriptions of new species are often insufficient, particularly in poorly known groups, and may not provide certainty as to the actual species the describer had before him. For this reason, every species has a unique "type" which can always be examined to determine to what species it belongs, making use of all the additional new information acquired since the original description. The word "type" for this exemplar, based on the essentialistic philosophy of the Linnaean period, is quite misleading because such a "type" is not particularly typical for the species, and the modern species description is not based exclusively on the type. Indeed, since every species and every population is variable, the description of the species must include the careful evaluation of this variability; in other words, it must be based on a large series of specimens.

The type of a species is a specimen; the type of a genus is a species (the type species); and the type of a family is a genus. The name of a family must be formed from the stem of the name of the type genus. The locality at which the type specimen of a species was collected is the type locality. This information is important in all polytypic species, that is, in species that consist of several geographic subspecies.

If several names are available for a taxon, the oldest one is ordinarily the valid name. However, it has happened, particularly in the early stages of taxonomy, that an older name was overlooked or was rejected for various reasons, and a junior name became the universally adopted name of the taxon. Information retrieval is severely handicapped when, at a much later period, the previously neglected older name is reinstated merely for the sake of priority. There are provisions in modern codes that state under what conditions such a prior name can be suppressed for the sake of the stability of nomenclature. The principle of priority is applied in zoological nomenclature only to the names of species, genera, and families, not to those of the higher taxa.[21]

The System of Organisms

Up to about the middle of the nineteenth century, organisms were classified into animals and plants. Anything not clearly an animal was placed with the plants. However, the closer study of fungi and micro-organisms made it clear that they had nothing in particular to do with plants but should be recognized as independent higher taxa. The most drastic revision of the classification of organisms resulted from the insight achieved in the 1930s that the Monera (prokaryotes), consisting of the bacteria and their relatives, were something entirely different from all other organisms (eukaryotes) with their nucleated cells.

From the origin of life (about 3.8 billion years ago) until about 1.8 billion years ago, only prokaryotes existed. They are now usually divided into two kingdoms, the Archaebacteria and the Eubacteria, mainly differing in their adaptations and in the structure of their ribosomes.[22] Around 1.8 billion years ago the first one-cellular eukaryotes originated, characterized by a membrane-enclosed nucleus with discrete chromosomes and by the possession of various cellular organelles. The latter evidently evolved through the inclusion of symbiotic prokaryotes. The exact details of the origin of this symbiosis, and in particular how the nucleus came into existence, are still controversial. The first fossil records of multicellular organisms appeared as recently as about 670 million years ago.

There are a number of possible ways of classifying the eukaryotes. Until recently, for the sake of convenience, the unicellular eukaryotes were usually combined into one taxon, the protists (Protista). Although it was fully understood that some of the protists (Protozoa) were closest to the animals, that others were closest to plants, and that still others were closest to the fungi, the traditional diagnostic criteria of plants and animals (possession of chlorophyll, mobility) often broke down at this level, and there was simply too much uncertainty about relationships to retain the convenient label "protists." New researches, particularly by Cavalier-Smith, which make use of previously neglected characters (for instance, the presence of certain membranes) and of molecular characteristics, have brought considerable clarification.

Although it may still be convenient to speak of unicellular eukaryotes as protists, a formal taxon Protista is no longer defensible. Whether to recognize 3 or 5 or 7 kingdoms for these protists is still being argued between lumpers and splitters.[23] For the nonspecialist it is probably convenient to recognize a smaller number. Thus the system of organisms might be divided into two empires and their respective kingdoms:

> Empire Prokaryota (Monera)
> > Kingdom Eubacteria
> > Kingdom Archaebacteria
> Empire Eukaryota
> > Kingdom Archezoa
> > Kingdom Protozoa
> > Kingdom Chromista
> > Kingdom Metaphyta (plants)
> > Kingdom Fungi
> > Kingdom Metazoa (animals)

CHAPTER EIGHT

"How?" Questions:
The Making of a New Individual

Every species consists of thousands, millions, or even billions of individuals. Many of them perish every day and are replaced by new ones. Although we usually think of sexual reproduction as the mechanism for generating new individuals, the simplest way to make a new individual is for an existing one to split into two. This is the normal way of reproduction in prokaryotes, in many protists and fungi, and even in some invertebrate phyla.

In addition to splitting, there are several other ways to reproduce without sex. A frequent pattern in some plants and invertebrates is to produce a new individual by budding. Somewhere on the body wall a bud originates which eventually breaks off and becomes a new individual. Vegetative reproduction, particularly through subterranean runners, is also frequent in plants. In some asexual organisms, new individuals develop from eggs alone—fertilization is not required. This process is called parthenogenesis. Aphids, planktonic crustaceans, and some other animals may alternate between parthenogenetic and sexual generations.

Most new individuals in higher organisms come into being exclusively through sexual reproduction, which involves many complex events in the production of eggs and sperm, mating of the two sexes, and the care of the developing embryo. Not surprisingly, this has produced one of the most protracted controversies in evolutionary biology: to explain the selective advantage of this reproductive strategy.

A female producing offspring through parthenogenesis has seemingly double the fertility of a female who wastes, so to speak, about half her descendants on males that are not capable of reproducing themselves. The ultimate explanation for the success of sexual reproduction is that it greatly increases the genetic variability of the offspring, and increased variability has multiple advantages in the struggle for survival—the reduction of vulnerability to diseases being only one of them.

Except for the workings of the brain, no other phenomenon in the living world is as miraculous and awe-inspiring as the development of a new adult from a fertilized egg. The history of our understanding of this process can be divided roughly into three periods. The first period, ranging from antiquity to about 1830, focused on describing the developing embryo. This period was particularly concerned with the relative contributions to the embryo made by the father and the mother. A second period began with the cell theory, when it was discovered that the vertebrate egg was a single cell and that the fertilizing element in the semen, the spermatozoon, was likewise a single cell. Investigators during this time were especially interested in the division of the fertilized egg into cells and in the eventual fate of each of these cells, that is, their contribution to the different structures and organs. By necessity, embryology had to be largely descriptive during these first two periods. The aim was to discover *what* happens.

During the third period, it became possible to investigate *how* development occurs—that is, the mechanisms that result in the formation of embryonic structures. Beginning early in the twentieth century, it was shown that development is controlled by specific genes and also that complex interactions occur between the parts of the embryo. Thus the behavior of developing cells was attributable not just to genes but also to the cellular environment in which these cells found themselves at different stages in development.

By necessity, the analysis of genes and gene-controlled biochemical processes had to be reductionist at the beginning, but it was soon realized that the genes interact with one another and with the cellular environment, much like musicians in an orchestra. The study of this well-orchestrated interaction of genes and cells during the making of an individual is currently the frontier of developmental biology. But this study could not begin until after centuries of careful descriptive work. Discovery was painfully slow.

The Beginnings of Developmental Biology

Diversity is the outstanding characteristic of the living world, and this is also true of developmental processes. Yet related organisms usually have similar developments. That the development of a chick in the incubated egg is a process akin to that of the mammalian embryo—also a vertebrate—was already vaguely perceived by the Egyptians, perhaps 1,000 years BC. But what little was previously known was completely eclipsed by Aristotle's great writings on descriptive and comparative embryology in animals. He established the field of reproductive biology, by discussing the nature of maleness and femaleness, the structure and function of the reproductive organs, viviparity (characterized by live births) versus oviparity (characterized by eggs that hatch outside the body), the form of copulation in different kinds of animals, the origin and characteristics of semen, and almost every other conceivable aspect of reproduction and development.

Indeed, Aristotle already faced two major problems in the field of reproduction that remained controversial until the very end of the nineteenth century. One is the theory of pangenesis (that every cell in the body contributes hereditary materials to the germ cells) and the other the debate over preformation versus epigenesis. It is almost inconceivable how this pioneer in the field of animal development could write an account of such completeness, based on such wide comparative observation and governed by such excellent judgment, that it was not surpassed until the nineteenth century.

Being human, however, Aristotle made a few mistakes. Although the females in all other groups of animals observed by Aristotle produced eggs, it apparently never occurred to him that mammalian females also might have eggs. Instead, he adopted the theory that the male semen gave form to the coagulate of the female's menstrual blood and that the mammalian embryo originated from this.[1]

Aristotle was long believed to have made a second error when he tried to explain the specificity of development, which so strongly impressed him. The egg of a frog invariably developed into a frog and not into a fish or chicken, as if it contained some information that would guide it toward its intended goal. This specificity induced Aristotle to postulate a "final cause" responsible for the unerring development of the egg to the adult stage. Only in our time was it realized that Aristotle's *eidos*, the seemingly metaphysical agent, is nothing else but what we now refer to as the genetic program, hence strictly explicable by physicochemical factors. The development of a fertilized egg is guided by a genetic program.[2]

Although reproduction and the development of embryos were surely of fascination through the centuries, the discipline of developmental biology did not make any real progress after Aristotle until Harvey in the seventeenth century carefully studied incubated hens' eggs with the naked eye and with the help of a simple lens. He clearly described a structure on the yolk membrane of a chicken egg as the spot from which the embryo originated. Harvey further demonstrated that there was no coagulated menstrual blood in the uterus of a mammal to serve as the female's contribution to the embryo, and he postulated the existence of a mammalian egg. Shortly afterward, the egg follicles were discovered in the ovary by Stensen and de Graaf, although the true mammalian ovum was not discovered until 1827 by Karl Ernst von Baer. It became clear that the ovary was the female equivalent of the male testis.

Much detail about the development of the chick was discovered in the years after Harvey, particularly through the use of early compound microscopes. First it was Malpighi, later Spallanzani, von Haller, and Caspar Friedrich Wolff, who greatly expanded our knowledge of the details of chick development. All these investigators, however, still tried to correlate the gradual development of the embryonic organs with Aristotle's physiological theories. This was the conceptual framework into which they attempted to squeeze their observations.

By contrast, nineteenth-century embryology was carried out in an entirely different spirit—one might almost say in a more truly scientific spirit. In all areas of functional biology, secure facts became the indispensable basis for sound theories. The three great representatives of early nineteenth-century embryology, Christian Pander, Heinrich Rathke, and von Baer, first carefully described their findings, mainly based on the chick, and only then theorized about them.[3] This included the recognition of the notochord, the neural tube, and most importantly, the three germ layers. These embryologists compared their findings in the chick with those in other vertebrates, and eventually even with the crayfish and other invertebrates.

The development of the chick (and the rather similar development of a frog), being readily available, has traditionally been considered the gold standard of embryology. Both are characteristic only of vertebrate development, however, while there is an endless number of separate developmental pathways in the other phyla of organisms.[4] The pattern of cleavage of the developing egg, in particular, may differ strikingly in different groups. When the nineteenth-century practitioners of experimental embryology compared the development of the vertebrates with that in tunicates, echinoderms, molluscs, coelenterates, and other invertebrate phyla, many differences became apparent. Most of the generalizations in the following pages apply mainly to the vertebrates.

The Impact of the Cell Theory

One of the numerous unifying contributions of the cell theory, proposed in the 1830s by Schwann and Schleiden, was to give new meaning to the terms eggs and semen, which up to that time had been rather formless concepts. Remak (1852) was the first person to demonstrate that the egg is a cell. But even after Leeuwenhoek in 1680 had discovered spermatozoa in the semen, it was widely held that they were merely parasites in the semen. Others declared them to be the carriers of the father's contribution to the embryo, but it was not realized that each spermatozoon is one cell, the male germ cell, until this was demonstrated by Kölliker (1841).

Curiously, up to about 1880 considerable uncertainty still remained as to the meaning of fertilization. For the physicalists, fertilization was merely the impulse or signal that initiated the cleavage divisions of the egg cell. This is how Miescher, the discoverer of DNA, interpreted fertilization as late as 1874. Eventually cytologists such as O. Hertwig and van Beneden showed that the spermatozoon brought far more to the egg than merely the command to start the first cleavage division; it also brought the nucleus of the male germ cell (gamete).

This nucleus, with its haploid set of male chromosomes, enters the egg cell. These chromosomes combine with the haploid set of female chromosomes of the egg cell, to form the diploid nucleus of the zygote. Fertilization, thus, not only restores diploidy but also combines in the offspring the genes of mother and father. The plant hybridizers, such as Koelreuter, had discovered this long before.

EPIGENESIS OR PREFORMATION?

But how can this apparent blob of "unformed" material of the zygote give rise to a chick or a frog or a fish? This puzzle led to a controversy in the seventeenth century that lasted until the twentieth century. Eventually two major hypotheses developed, both of them based on good arguments and both of them now known to have been partly right and partly wrong. These were the hypotheses of preformation and epigenesis.

The preformationists derived their hypothesis from the observation that a fertilized egg unerringly produces the adult of the species that had produced the egg. From this they concluded that at fertilization a miniature form of the future organism is already present in the egg or sperm and all development is merely the unfolding—which they called "evolution"—of this original form. This theory was strengthened by the claim of the earliest pronounced preformationist, Malpighi, that when he looked at a fertilized hen's egg he was able to see the earliest stages of development, indicating to him that the form of the future organism was already preformed in the egg.

The logical extension of the concept of preformation was the assumption that not only was an organism preformed but that in the preformed organism all of its descendants had to be present. This extension of preformation was called the theory of *emboîtement*. The further question arose as to the location of the preformed individual: Was it in the egg, as claimed by the ovists, or was it in the sperm, as claimed by the animalculists? Numerous descriptions and illustrations in the literature during this period showed a little man (homunculus) enclosed in the spermatozoon.

Koelreuter's hybridization experiments (1760) with plants clearly refuted both preformationist theories by showing that hybrids were equally determined by both father and mother. There could not have been a preformed adult of the species in a germ cell of only one of the parents. Perhaps because his experiments were done with plants, this decisive disproof of preformation was long ignored. But so was the intermediacy of mules and other animal hybrids. Equally ignored were the findings of regeneration, which showed that when major parts of certain organisms were removed, as in the hydra or in certain amphibians and reptiles, they could be regenerated by what was essentially an epigenetic process.

The epigenesists, who opposed the preformationists, thought that development started from an entirely unformed mass that was given form by some extraneous force, a *vis essentialis,* as it was called by C. F. Wolff.[5] But epigenesis could not explain why the eggs of a chicken produced chickens and those of a frog, frogs, nor could it explain the differentiation of tissues and embryonic structures during ontogeny. Moreover, belief in epigenesis meant that every species had to have its own *vis essentialis,* something quite different from the universal forces described by the physicists, such as gravity. None of the epigenesists could explain what the *vis essentialis* was and why it was so specific.

Nevertheless, epigenesis won out in the controversy, particularly after improved microscopic techniques could not find any trace of a preformed body in the newly fertilized egg. But the ultimate solution to this puzzle was not found until the twentieth century. The first step came from the field of genetics, which distinguished between a genotype (the genetic constitution of an individual) and a phenotype (the totality of the observable characteristics of an individual) and showed that during development the genotype, by containing the genes for becoming a chick, could control the production of a chick phenotype. By thus providing the information for development, the genotype is the preformed element. But by directing the epigenetic development of the seemingly formless mass of the egg, it also played the role of the *vis essentialis* of the epigenesists.

Finally, molecular biology removed the last unknown by showing that the genetic DNA program of the zygote was this *vis essentialis*. The introduction of the concept of a genetic program terminated the old controversy. The answer was thus, in a way, a synthesis of epigenesis and preformation. The process of development, the unfolding phenotype, is epigenetic. However, development is also preformationist because the zygote contains an inherited genetic program that largely determines the phenotype.

That the ultimate answer in a long-lasting controversy combines elements of the two opposing camps is typical in biology. Opponents are like the proverbial blind men touching different parts of an elephant. They have part of the truth, but they make erroneous extrapolations from these partial truths. The final answer is achieved by eliminating the errors and combining the valid portions of the various opposing theories.

DIFFERENTIATION, THE DIVERGENCE OF DEVELOPING CELLS

One of the most wonderful, and for a long time totally inexplicable, aspects of development is the gradual differentiation among the cells descending from the single cell of the zygote. How does a nerve cell become so different from the cells of the intestinal tract?

The problem of cell differentiation became even more puzzling in the 1870s and 1880s, when it was finally realized that genetic determination resided in the cell nucleus, and more specifically in the chromosomes. If the nucleus of every cell in the body contained the same genetic determinants, as was claimed by Weismann, how could cells become so different during the course of development?

The simplest solution was to assume that during mitotic cell division, when the chromosomes divided, a somewhat different assortment of chromosomes with different genetic elements would go to the two daughter cells, and cell differentiation would depend on the specific genetic elements that the cell received. This theory of unequal cell division was, no doubt, the majority opinion from the 1880s until at least 1900. But if this were true, then the elaborateness of mitosis, as observed by the cytologists, would make no sense. Roux (1883) asked quite rightly why the nucleus did not simply divide along its equatorial plane with both half-nuclei becoming the nuclei of the two daughter cells. What is the sense of this elaborate mechanism converting during mitosis each chromosome into one single very long string of chromatin? This makes sense, as Roux pointed out, only if the nucleus consists of highly heterogeneous material, perhaps of uniquely different particles. In that case, an equal distribution of these particles into the two daughter cells is possible only if these particles are strung up on a single thread, so to speak, and then this thread is sliced longitudinally. This would guarantee a completely equal distribution of the heterogeneous contents of the nucleus to the two daughter cells.

We now know that Roux's theory was essentially correct and was a most brilliant deduction from his observations of mitosis. Alas, it seemed to be refuted by some observations made in the ensuing years, and Roux himself eventually gave up his valid original theory and accepted instead unequal mitotic division. The reasons for this conversion were studies which showed that after the earliest cleavage divisions, the descendant cells in some organisms were exceedingly different and gave rise to very different organ systems. How could this possibly happen if the genetic elements were divided equally?

Other findings deepened the mystery. Experiments by Roux, Driesch, Morgan, and Wilson showed that the early cleavage cells of different animal groups had different "potencies." Cleavage cells in an ascidian, when separated, produced a lineage of descendant cells that would have the same properties as if they had not been separated; the two cells produced by the first cleavage division would produce two half-ascidian larvae. This mode of differentiation has been referred to as *mosaic* or *determinate development*. But when the two cells of the first cleavage division of a sea urchin are separated, these two cells eventually produce two near normal larvae, although of reduced size. This very different mode of differentiation came to be called *regulative development*. To complicate matters even further, development in many groups turned out to be somewhat intermediate between these two modes.

The more the details of development in different organisms were studied, the more difficult it became to establish clear-cut general principles. The processes in one kind of organism often turned out to be different from those in another one. Some developing cells seemed to be impervious to influences from their cellular environment; others could be completely reprogrammed by it. Some cells stayed right in the tissue in which they were first laid down; others went on more or less extensive migrations within the embryo. At the conclusion of numerous experiments, the nature of the relation between genotype, on the one hand, and differentiation of cleavage cells, on the other, remained long a riddle.[6]

Eventually, particularly through the contributions of twentieth-century molecular biology, it was realized that all cells undergo a process of differentiation and that at any particular time only a small fraction of the genes in the nucleus of a given cell are active. Regulatory mechanisms turn on or turn off a given gene, depending on whether its gene product is needed in that cell at that time. The timing of this regulatory activity is in part programmed in the genotype and in part determined by neighboring cells. Even so sophisticated a biologist as Weismann was unable to conceive of the possibility of such an elaborate capacity of the genotype, and he opted, therefore, for the erroneous solution of unequal nuclear division. Even today, it is poorly understood how the regulatory genes "know," that is, sense, when to activate other genes.

It was further discovered that control of the early cell divisions in many zygotes, particularly yolk-rich ones, is entirely due to maternal factors in the cytoplasm. This is what had misled Roux. Only after the earliest stages of development have been completed do the nuclear genes of the new zygote take over. How the ovary determines what material to place in the different parts of the egg yolk and how it transfers this material appropriately is still a deep mystery.

In the nematode *Caenorhabditis,* for example, the founder cell of several different cell lineages is provided with a specific sector of egg cytoplasm which, it is assumed, contains regulatory factors of maternal origin. By contrast, in taxa with regulative development, as in the vertebrates, there are no fixed early cell lineages; there is extensive cell migration; and induction (the influence of existing tissues on the development of other tissues) largely determines the specificity of cells. Profound differences in the pathways of differentiation can be found not only between nematodes and vertebrates but even between species of more closely related phyla—example, between chordates (including vertebrates) and echinoderms. There is a great variety of developmental patterns, and some proceed independently of all environmental influences while others are greatly affected by them.

FORMATION OF THE GERM LAYERS

The students of development in the eighteenth century, working with a primitive methodology, thought that the heart was the first structure to appear in ontogeny and that other organs appeared when they were functionally needed by the developing embryo. C. F. Wolff, Pander, and von Baer showed, however, that this was not at all the case.

Rather, through the first 8 to 12 cleavage divisions of a frog egg a ball of cells is formed, the so-called blastula. Into the hollow of this blastula, part of the outer layer of cells "invaginates," resulting in the double-layered gastrula. Finally, a median layer develops (by a number of different processes), called the mesoderm. The cells that will form the three germ layers form the outside of the blastula. Those that will become the ectoderm are the upper hemisphere; those in the equatorial region are the mesoderm. Most of the ventral hemisphere will become endoderm. Pander (1817) first demonstrated the existence of these three cell layers in the developing chick, and a few years later von Baer (1828) showed that the production of three germ layers characterized development in all classes of vertebrates. Each germ layer gave rise to a particular set of organ systems: the ectoderm to the skin and nervous system, the endoderm to the intestinal system, and the mesoderm to muscles, connective tissue, and the blood system.

After the 1830s, the application of the cell theory increased investigators' understanding of the development of the germ layers. It was soon realized that an ectoderm and endoderm also exist in all groups of invertebrates, particularly the coelenterates. Also, the formation of the germ layers was the same in all groups of organisms, consisting of an invagination of the ectoderm of the blastula resulting in the formation of the gastrula.[7]

By the end of the 1870s, considerable doubts had arisen as to whether the same germ layers gave rise in all organisms to the same structures, and, in particular, about the relation of the mesoderm to the other two germ layers. Experiments with regeneration, treatment with various chemicals, and the analysis of pathologies all indicated that germ layers could adopt roles that were different from their normal one.

A new era in the study of the potential of the germ layers began when surgical methods were introduced into experimental embryology, particularly transplantation experiments. They showed that when pieces of a germ layer were transplanted to a new location in the embryo or cultured in tissue culture, development was often different from that at the normal location. For instance, isolated ectoderm failed to differentiate nerve tissue in tissue culture; it formed only epidermis when deprived of the influence of cells of the other layers. When tissues of early amphibian embryos were implanted into the abdominal cavity, ectodermal as well as endodermal tissues were able to differentiate structures normally produced by the other germ layers. The result of all these experiments was that the doctrine of the absolute specificity of the germ layers, widely adopted in the last century, could no longer be upheld. The germ layers seemed to have a normal potentiality when in their usual relation to other germ layers or cell complexes, but revealed additional potentialities when the normal relationship was disturbed.

Moreover, it was discovered that the germ layers do not retain their integrity throughout development. Instead, many embryonic cells undertake long migrations. The mesoderm, for instance, may be formed from cells that migrate from the ectoderm *or* endoderm. Pigment cells and neurons in vertebrate embryos undertake long migrations from their place of origin in the neural crest. In some cases, the migrating cells definitely are attracted by chemical stimuli emanating from the target area, in a process called induction.

INDUCTION

Around 1900, the distinction, first made by Roux, between tissues or structures that seem to develop strictly according to a fixed genetic program (determinate development) and others that are affected by adjacent tissues or structures (regulative development) eventually led to a new concept in experimental embryology called "induction." This term refers to all cases in which one tissue affects the subsequent development of another tissue.

The phenomenon was first clearly demonstrated by Spemann (1901) for the eye of the frog embryo. The lens is formed by the lens ectoderm, and yet it fails to develop if the underlying mesodermal tissue (eye anlage) is destroyed or removed. It was said that the eye anlage induces the formation of the lens. Spemann tested his findings by transplanting the eye anlage to other parts of the body, to see whether the ectoderm of different body regions had the same lens-forming capacity. And indeed it had. Finally, he removed the local ectoderm of the eye region and replaced it by ectoderm from other parts of the body, and again a lens was formed. Subsequently, other authors obtained different results, primarily when working with other species of frogs. Sometimes there was "free-lens development" even after the eye anlage had been removed. Spemann finally concluded that a large region of head ectoderm did indeed possess a lens-forming predisposition.

In another series of transplantation experiments, Spemann showed that a portion of the dorsal blastopore lip induced neural tube tissue in the roof of the primitive gut. He hypothesized that an "organizer" was responsible for this effect. This paper—co-authored with Hilde Mangold, who had done most of the technical work, and published in 1924—caused quite a sensation and resulted in almost feverish activity among the experimental embryologists. Eventually it was shown that even "dead" organizers, and in fact even inorganic substances, are sometimes able to induce neural tube formation.

Spemann himself and many others in the field either stopped working altogether or turned to other problems, and yet it is now clear that he had been on the right track. Recently a protein was isolated that seems to have the capacity to induce neuronal tissue. A review of all experiments in the field led Spemann later to see induction as a complex interplay between the inducing and induced tissue.[8]

Regardless of the nature of the chemical signal sent by the inducing tissue to the induced one, it is well established that induction plays an important role in the development of organisms with regulative development (such as the vertebrates). The study of the interaction of cells and tissues during ontogeny, in particular the position-dependent behavior of cells, has now become an independent field of biology (topobiology), in which the properties of cell membranes are singled out for special analysis. It has become quite clear that the interaction of cells and tissues plays an important role in the development of nearly all organisms, except perhaps in a few with strictly determinate development.

RECAPITULATION

Naturalists all the way back to Meckel-Serrès and von Baer have been interested in the evolutionary implications of development. In the mid-1820s Rathke discovered the gill slits and pouches of the embryonic birds and mammals—an observation which fitted excellently into the thinking of the period of the "great chain of being" (scala naturae). If adult organisms could be arranged in a series of ever greater perfection, why should not their embryos go through an equivalent series of stages, reflecting the preceding archetypes of less advanced perfection? Surely, the gill slits indicated a fish stage, and still earlier embryonic stages represented recapitulations of still more primitive types.

Thus, the recapitulation theory was born, also referred to as the Meckel-Serrès law: organisms recapitulate during their ontogeny the phylogenetic stages through which their ancestors had passed. Evolutionary thinking in the pre-Darwinian period was still rather confused, but recapitulation fit in with a widespread idea that organisms "higher" in the scale of being went through earlier phylogenetic stages during their ontogeny.

Von Baer, even though confirming the similarity of some ontogenetic stages with those of "lower" types, categorically rejected the evolutionary interpretation. For him the earlier stages merely were simpler, more homogeneous and the later stages more specialized, more heterogeneous; all ontogeny was a move from simple to more complex (this was designated "von Baer's law"). Teleological interpretations were quite acceptable to von Baer, but anything like Darwin's theory of common descent was not.

The situation was different with Ernst Haeckel. Haeckel, more than anyone else, emphasized a recapitulationary aspect of development, proposing that the gastrula stage corresponded to the evolution of the invertebrates and that later stages of development corresponded to the evolution of "types" of "higher" organisms. Soon after the publication of Darwin's *Origin*, Haeckel proclaimed "the fundamental biogenetic law" as "Ontogeny recapitulates phylogeny." At once this raised enormous interest in comparative embryology, and students of ontogeny thought that they found confirmation of Haeckel's claims wherever they looked. For a few years in the late nineteenth century, embryology became the search for common ancestors with the help of evidence from recapitulation.

But on the whole embryologists have tended to reject the theory of recapitulation, particularly in its more extreme versions, in favor of von Baer's law. The reason for this choice was largely theoretical. They could not think of any convincing cause why an embryo should pass through ancestral stages, and they felt more comfortable with a progression from simple to complex, as claimed by von Baer. Indeed, embryos are usually simpler, less differentiated than the resulting adults. However, the supporters of von Baer neglected the fact that gill arches and other manifestations of recapitulation are never simpler than the resulting development. Von Baer's law merely swept recapitulation under the rug; it did not explain it.

Developmental Genetics

In the last quarter of the nineteenth century, development was also studied by a branch of biology that eventually came to be called genetics. But this new field was not homogeneous. The students of inheritance soon realized that their field consisted of two branches, one later called transmission genetics, the other one developmental or physiological genetics. Mendelian genetics, which dealt with the mode of transmission of the genetic factors from one generation to the next, was pure transmission genetics. Developmental genetics, on the other hand, dealt with the activity of these factors in organisms during ontogeny. The failure of some biologists, such as Weismann, to separate these two aspects of genetics was responsible for much of the early misunderstandings. The achievement of T. H. Morgan was to clearly separate the two and to confine himself strictly to an elucidation of transmission genetics.

In the same period other authors concentrated on developmental genetics, a field in which Richard Goldschmidt (1938) produced the first major text. Much of what was stated in this field at that time was pure speculation, and it was not until after the rise of molecular biology that developmental genetics began to mature. Yet earlier publications, such as those of Waddington and Schmalhausen, had already outlined most of the problems that are the object of modern research.

A new era in developmental genetics was opened when Avery (1944) demonstrated that DNA was the carrier of the genetic information. DNA controls the production of the proteins of which an organism is composed. Development, then, is the elaboration of different kinds of proteins during ontogeny and of the very specific combination of proteins that are characteristic of the different organ systems. Although the founders of modern genetics were fully aware of the connection between genes and development, they did not succeed—indeed they did not even seriously attempt—to produce a synthesis of genetics and development.

The emphasis in classical genetics was on individual genes. But at that time a given gene's contribution to development could be determined only by the study of mutations, particularly deleterious or even lethal mutations. There was no way to study the contribution to development of a normal (or, as it was called, "wild type") gene. Indeed, the analysis of deleterious genes was the preferred method of developmental genetics from the 1930s on. It produced modest results, very often pinpointing the particular tissue or even germ layer involved in the mutation. The analysis also showed that most mutations consisted of a failure to produce a needed gene product, but it failed to help in determining the biochemical nature of the deficiency.

Even though the chemical nature of the gene product remained unknown, these studies clearly demonstrated that a given gene is usually active during development only in particular tissues and at particular stages of the development. On the basis of this recognition, one could describe development as an ordered sequence of gene expressions.

THE IMPACT OF MOLECULAR BIOLOGY

The realization, provided by molecular biology, that the gene is not a protein and does not itself form one of the building blocks of the developing embryo, but that the genotype is simply the set of instructions needed for the construction of the embryo, had a profound impact on the methodology and conceptualization of developmental genetics. When the details of gene action began to be elucidated in the 1960s and 70s, it became evident why our previous explanatory schemes had fallen short.

Not only are genes composite, consisting of exons that are transcribed and introns that are excised prior to protein synthesis, but in addition to the enzyme-producing structural genes there are regulatory genes and flanking sequences. It finally became clear, as had been cautiously suggested since the 1880s, that a gene can be turned on and off whenever its product was needed. Furthermore, the molecular revolution helped us appreciate the fact that cells are characterized by the proteins they produce.[9]

The entire system, from the nuclear DNA through messenger RNAs to polypeptides and proteins, and the continuous interaction of this whole apparatus with its cellular environment, turned out to be far more complex than had been realized before. The ideal achievement of developmental biology would be to discover every last gene involved in development, to determine each gene's exact contribution, including the chemical nature of the relevant gene product and the role this molecule plays in development, and to analyze the regulatory machinery which controls the timing of the activity of each gene. Amazingly, developmental scientists are well on their way toward this goal in certain organisms.

The greatest progress has been made in those with rigidly determinate development, such as nematodes and *Drosophila*. In the nematode *Caenorhabditis elegans*, for instance, more than 100 genes with over 1,000 mutations have been mapped. Furthermore, the DNA of many of these genes has been sequenced, and the exact sequence of base pairs has been established. The adult nematode has a fixed number of 810 nongonadal cells, and through a study of cell lineages it has been possible to determine which organs are derived from which cells of the early cleavage divisions.

Drosophila (fruit fly), another organism with determinate development, has some disadvantages as a case study, such as the much larger number of genes, but this is more than compensated for by genetic and morphological advantages. First of all, when the modern developmental studies began there was already a huge inventory of *Drosophila* mutations available. Moreover, their position on the chromosomes had been determined. Also, *Drosophila's* giant salivary chromosomes often allow the nature of the mutations to be elucidated.

But most importantly, *Drosophila* is a metameric organism, and through genetic analysis investigators can determine which genes contribute to the development of which segment. There are five head segments, three thoracic segments, and eight to eleven abdominal segments; numerous genes are now known to affect either particular segments or groups of segments. To a large extent it has been discovered which of these genes do what. Particularly interesting is a comparison of the effects of different alleles (versions of a gene) at the same gene locus.

Far less progress has been made in the genetic analysis of organisms with regulative development, such as vertebrates. In these species, cells are not yet committed until the 16 to 32 cell stage of development. Perhaps the greatest contribution to the understanding of development in humans has been made by the study of human genetic diseases, that is, of mutations resulting in deleterious changes in the phenotype. This has allowed investigators to assign a high percentage of mutations to particular chromosomes. No doubt, through the human genome project, all mutations will eventually be localized. But considering the regulative nature of development, the frequency of induction, and the extensive amount of migration of certain cell complexes, it will often be difficult to establish a one-to-one relation between specific genes and specific aspects of phenotypic development. The developmental systems of organisms with regulative development are considerably more complex than those of species with determinate development. One may have to be satisfied with generalized conclusions.

One of the most exciting developments in molecular embryology has been the discovery that certain clusters of genes are widely distributed among only distantly related groups of animals. These so-called *Hox* genes were first discovered in *Drosophila* but, through sequence analysis, were also found in the mouse, in an amphibian, in a nematode, and in other animals. There are, for example, four homologous clusters of *Hox* genes in vertebrates. These clusters seem to encode relative position within the organism rather than any specific structure. Homologous *Hox* genes were also discovered in most phyla of invertebrates, from the coelenterates and flatworms to arthropods, molluscs, and echinoderms. A certain number of the *Hox* cluster genes, together with a number of other development-controlling genes, are so widely distributed among the animal phyla that it has been suggested by Slack et al. (1993:491) that this set of genes (they call it the "zootype") reflects part of the genotype of the ancestral metazoan. Unquestionably, this assembly of genes is of great phylogenetic age. Which of these genes are also found in the protist ancestors of the animals is still unknown.

Development and Evolutionary Biology

For a while, when most geneticists thought that evolution was merely a change in gene frequencies, the role of development in macroevolutionary changes was neglected. In recent years, particularly following the developmental biologists' rather reluctant acceptance of Darwinism, legitimate stress has again been placed on this very interesting aspect of development.

The individual, the principal target of selection, is the product of the interaction during development of all its genes with one another and with the environment, and this interaction sets narrow limits on permissible evolutionary changes. This fact is shown by the phenotypic uniformity of most species. Any deviation from the standard morpho-type of the species will be eliminated through stabilizing or normal-izing selection (see Chapter 9).[10] A study of these developmental constraints on evolution has become one of the major areas of interest in modern developmental biology.

Different genes and sets of genes are active at different stages in the development of the zygote. Developmental biologists have long be-lieved that the genes active near the end of development are the ones acquired last during phylogeny and, conversely, that the genes active earliest in development are the "oldest" genes of an organism. Any change in a recent gene, so it was believed, would effect only a minor change in the phenotype, let us say, by changing the degree of sexual dimorphism or by affecting a behavioral component of an isolating mechanism, while a mutation of one of the early genes may lead to a fundamental change in the whole process of development and therefore most likely be deleterious.

Many objections have been raised against too literal an interpreta-tion of this concept, and yet there are numerous observations suggest-ing that it is perhaps valid in principle. If so, it would explain many evolutionary phenomena, such as the exuberance in the production of new structural types in the Precambrian and early Cambrian when the metazoan genotype was still young, in contrast with the relative stability of structural types ever since. It would explain also, for instance, why evolutionary innovations are often due to a change of function of a structure that had been gradually acquired step by step for a different function. Such a shift of function has the advantage that it requires only a minimal restructuring of the genotype.

The realization that every individual is a developmental system which reacts to selection more or less as an integrated system explains also two evolutionary phenomena that long puzzled developmentalists. The first is the existence of vestigial structures. Most genes and groups of genes have widespread effects, and even when one of the phenotypic manifestations of such a group of genes, let us say the presence of a vestigial digit, is no longer supported by natural selection, this vestigial character will not be lost as long as the controlling genes still have other functions, let us say in the maintenance of the other digits. If so, it will be maintained by natural selection. The second evolutionary phenomenon is recapitulation.

RECAPITULATION RECONSIDERED

In order to explain recapitulation in terms acceptable to a modern biologist, one must start on a new basis. The Meckel-Serrès principle was proposed at a time when idealistic morphology was ruling. Haeckel and other proponents of recapitulation knew perfectly well that no bird or mammal went through an embryonic stage that was exactly like a fish. They did not claim, as they were accused by their opponents, that the embryonic stages of a mammal or bird were exactly the same as the "adult" stages of amphibians or fish. Rather, they claimed that the embryonic stages resembled the "permanent" stages of their ancestors. What they meant by "permanent" was that the earlier ontogenetic stages represented the antecedent archetypes.[11] Indeed, these recapitulationists pointed out that the earlier ontogenetic stages often had advanced further evolutionarily than the adult stages. This is particularly true for organisms in which the larval stages were adapted for special modes of living, as for instance the larvae of some marine organisms and parasites.

In evaluating the theory of recapitulation one must distinguish two sets of questions: (1) Do ontogenetic stages sometimes resemble those of ancestral types? That is, does "recapitulation" actually occur? (2) If so, why does it occur? Why is there such a permanence of the ancestral ontogenetic stages? The answer to the first question is yes. But in the case of the second question, one is justified in asking: Why does a mammal not develop the neck region directly instead of roundabout through the gill arch stage? The answer is that the development of the phenotype is not strictly, exclusively, and directly controlled by genes but by the interaction between the genotype of the developing cells and their cellular environment. At any stage of ontogeny, the next stage of development is controlled both by the genetic program of the genotype and by a "somatic program" consisting of the embryo at this stage. To apply this, for instance, to the gill arch problem, it means that the gill arch system is the somatic program for the subsequent development of the avian and mammalian neck region (Mayr 1994). In spite of the new term "somatic program," this interpretation is more than one hundred years old. It has long been one of the fundamental ideas of developmental biology that any stage of development is in part controlled by the previous stages. There is thus nothing mysterious about recapitulation except that it must be divorced from the typological thinking of idealistic morphology.

Despite the many complexities and variations from group to group of organisms, the early development of animals, as reflected in the formation and development of the germ layers (gastrulation), shows great similarity throughout all phyla. Somehow I cannot suppress the feeling that this stage may represent the recapitulation of an ancestral condition. Haeckel's extravagant theories have made this thought highly unpopular, but even a hard-nosed look at the facts does not lead me to a different and superior interpretation.

HOW EVOLUTIONARY ADVANCES OCCUR

The developmental system is so tightly knit that biologists often speak of the "cohesion" of the genotype. For evolutionists the problem is how this cohesion developed, and how it is broken to make major new evolutionary advances possible.

According to a model I proposed in 1954, evolution progresses rather slowly in large, populous species, while most rapid evolutionary changes occur in small, peripherally isolated founder populations.[12] Expressed in terms of development, this suggests that large populous species are developmentally stable, while small founder populations may lack this stability, enabling them to shift quickly to a new phenotype through rapid genetic restructuring. Eldredge and Gould (1972), using the phrase "punctuated equilibria," accepted this model and proposed that the developmental stasis of populous species may last through millions of years. Subsequent research has confirmed that this is indeed true for many species. This model quite clearly stresses the importance of development in macroevolution. However, it does not explain why the genotypes of certain species are highly stable while those of other species can undergo rapid evolutionary change. This difference is unexplained even today.

This model is almost the exact opposite of the one proposed by Fisher and Haldane in the early 1930s. According to their view, the rate of evolutionary change is correlated with the amount of genetic variance in a population or species, and therefore the larger and more populous a species is, the more rapidly it evolves. All subsequent researches have clearly refuted the Fisher-Haldane thesis. My opposing interpretation is that the more populous a species is, the more epistatic interactions occur and the longer it will take for a new mutation or recombination to spread through the entire species and therefore the slower evolution will proceed. A founder population, with less concealed variation because of having fewer individuals, can more readily shift to another genotype, or, to use another metaphor, to another adaptive peak. Change in evolutionary rates in populations and species, caused by mutation or genetic recombination, is called "heterochrony."

It is now well understood that considerable genetic variation exists at every stage in the hierarchy of developmental processes. Milkman (1961) showed beautifully how much cryptic genetic variation there can be in a natural population for the expression of a single phenotypic character. Such variation permits natural selection to affect developmental processes. Many morphological properties are evidently closely correlated with physiological processes. Selection pressure on these pleiotropic physiological processes is often responsible for otherwise inexplicable morphological changes.

By comparing changes in developmental processes in different geographical races and closely related species, developmental biologists should be able to show what kind of developmental changes are possible in close relatives and what others are not. Unfortunately for studies such as these, developmental biologists' traditional methodology has permitted, if not actually favored, typological thinking. Darwinian population thinking was rarely required in their researches. A few of them, such as Waddington, appreciated the existence of variation, but the gradual acceptance of population thinking among developmental biologists has been a slow process. Developmental biologists in the past have tended to go for their analysis to model systems in the laboratory—the chick, the frog, *Drosophila*—and have gone directly from the phenotype to the gene level. Until recently they failed to take advantage of the pathway that is truly responsible for the initiation of most macroevolutionary events, namely, geographic variation.

Yet in no other branch of biology are the different explanatory aspects of the life sciences represented in such exemplary fashion as in developmental biology. This discipline is highly analytical (often misleadingly called reductionist), with the goal of determining the contribution that each gene makes to the developmental process. At the same time it is conspicuously holistic, since viable development depends on the influence of the organism as a whole, reflected by the interaction among genes and tissues. The decoding of the genetic program represents the proximate causation of ontogenetic processes, while the contents of the genetic program are the result of ultimate (evolutionary) causations. It is this richness of factors and causations that is the fascination and beauty of the living world.[13]

CHAPTER NINE

"Why?" Questions: The Evolution of Organisms

In the Middle Ages and almost up to Darwin's time, the world was believed to be constant and of short duration. But the credibility of this Christian worldview had already been weakened in some quarters by a series of scientific developments. The first of these was the Copernican Revolution, which had removed the earth and its human inhabitants from the center of the cosmos and in the process had demonstrated that not every statement in the Bible had to be interpreted literally. Second, the researches of the geologists had revealed the great age of the earth, and, third, the discovery of extinct fossil faunas had refuted the theory that the earth's biota was unchanged since the Creation.

Despite this and much more evidence which undermined the theory of a constant world of short duration (and even though doubts were expressed in the writings of Buffon, Blumenbach, Kant, Hutton, and Lyell, as well as in Lamarck's full-fledged theory of gradual change), the more or less biblical worldview still prevailed up to 1859. It was popular not only among lay people but also among the majority of naturalists and philosophers. A long series of developments was required before evolutionism—which posits an ever-changing world of long duration—was fully established. It may seem strange to us today, but the concept of evolution was alien to the Western world.

The Manifold Meanings of "Evolution"

The word "evolution" was introduced into science by Charles Bonnet for the preformational theory of embryonic development (see Chapter 8), but developmental biology no longer uses the word in this sense. Evolution has also been used for three concepts of the history of life on earth and is still used for one of them.

Transmutational evolution (or transmutationism) refers to the sudden origin of a new type of individual through a major mutation or saltation; this individual becomes the progenitor of a new species through his descendants. Saltational ideas, although not under the designation evolution, had been proposed from the Greeks to Maupertuis (1750). Even after the publication of Darwin's *Origin*, saltational theories were adopted by many evolutionists—including Darwin's friend T. H. Huxley—who could not accept the concept of natural selection.

Transformational evolution, by contrast, refers to the gradual change of an object, such as the development of a fertilized egg into an adult. All stars experience transformational evolution, as from a yellow to a red star. Nearly all changes in the inanimate world, such as the rise of a mountain range owing to tectonic forces or its subsequent destruction by erosion, are of this nature, if they are directional at all. As for the animate world, Lamarck's theory of evolution, which preceded Darwin's, was transformational. According to Lamarck, evolution consists of the origin by spontaneous generation of a simple new organism, an infusorian, and its gradual change into a higher, more perfect species. Lamarck's theory of transformational evolution, as presented in his *Philosophie Zoologique* (1809), although at one time widely adopted, has been replaced in most parts of the world by Darwin's theory.

Variational evolution is the concept represented by Darwin's theory of evolution through natural selection. According to this theory, an enormous amount of genetic variation is produced in every generation, but only a few survivors of the vast number of offspring will themselves reproduce. Individuals that are best adapted to the environment have the highest probability of surviving and producing the next generation.

Owing to (1) the continuing selection (or differential survival) of genotypes best able to cope with the changes of the environment, (2) competition among the new genotypes of the population, and (3) stochastic (chance-based) processes affecting the frequency of genes, there will be a continuous change in the composition of every population, and this change is called evolution. Since all changes take place in populations of genetically unique individuals, evolution is by necessity gradual and continuous, as populations are genetically re-structured.

In his earlier writings (the Notebooks), Darwin had been well aware of two evolutionary dimensions: time and space. Transformation in time (phyletic evolution) deals with changes in adaptedness, as when a given species acquires new characteristics. But this concept alone can never explain the extraordinary diversification of organic life because it does not allow for the number of species to increase. Transformation in space (speciation and multiplication of lineages) deals with the establishment of multiple new populations outside the range of the parental population, and with their change into new species and eventually into higher taxa. This multiplication of species is called speciation.

Lamarck had had absolutely nothing to say about the geographical (speciational) aspect of evolution, and indeed, being a transformation-ist and having accepted spontaneous generation, he seems not to have been aware that the question "How do species multiply?" needed to be asked. Even Darwin neglected the subject in his later writings. Paleontologists in Darwin's time and for decades afterward continued to adhere to the idea that phyletic evolution was the only kind of evolution that mattered. Only in the 1930s and 40s was it finally emphasized, in the works of Dobzhansky and Mayr, that evolution is as much transformation in space as it is transformation in time, and that the origin of organic diversity through speciation was as important a concern of evolutionary biology as are adaptive changes within a lineage.

Darwin's *Origin of Species* established five major theories relating to different aspects of variational evolution: (1) that organisms steadily evolve over time (this we might designate as the theory of evolution as such), (2) that different kinds of organisms descended from a common ancestor (the theory of common descent), (3) that species multiply over time (the theory of the multiplication of species, or speciation), (4) that evolution takes place through the gradual change of populations (the theory of gradualism), (5) and that the mechanism of evolution is the competition among vast numbers of unique individuals for limited resources, which leads to differences in survival and reproduction (the theory of natural selection).

Darwin's Theory of Evolution as Such

In the *Origin* Darwin presented a great deal of evidence in favor of the theory that animals evolve over time. In the following decades biologists searched for and found abundant favorable—and no contrary—evidence that evolution as such has occurred. In the more than a century and a quarter since Darwin's time this evidence has become so overwhelming that biologists no longer speak of evolution as a theory but consider it a fact—as well-established as the fact that the earth rotates around the sun and that the earth is round and not flat. As Dobzhansky has said: "Nothing in biology makes sense except in the light of evolution." Considering evolution to be an established fact, no evolutionist any longer wastes time looking for further evidence. It is only when refuting creationists that one may bother to assemble the powerful evidence that has accumulated in the last 130 years proving evolution.

THE ORIGIN OF LIFE

One objection to the theory of evolution made by Darwin's early opponents was that though he may have explained the derivation of organisms from other organisms, he had not explained the origin of life itself from inanimate matter. The researches of Louis Pasteur and others demonstrating the impossibility of spontaneous generation in an oxygen-rich atmosphere seemed to strongly support the idea that life cannot arise from natural causes but requires some supernatural origin, a Creator.

It has since been discovered that, unlike today, there was no oxygen (or only traces of it) in the early atmosphere of the earth, when life originated.[1] Experiments carried out by Stanley Miller (1953) showed that electrical discharges sent through a gaseous mixture of methane, ammonium, hydrogen, and water vapor in a flask would result in the production of amino acids, urea, and other organic molecules. Such organic molecules could have accumulated when our atmosphere was devoid of oxygen, and, indeed, similar molecules have since been found in meteorites and in interstellar space.

NB

There are now numerous hypotheses to explain how life, particularly proteins and RNA, might have emerged from a combination of these organic molecules. Several of these prebiotic scenarios are quite convincing, but in the absence of any chemical fossils of the intermediate stages we may never be able to prove which of the scenarios is the right one. It would seem that the first organisms were heterotrophic, that is, they utilized prebiotically produced organic compounds available in the environment. The organisms had to build the larger macromolecules such as proteins and nucleic acids, but they did not have to synthesize *de novo* the amino acids, purines, pyrimidines, and sugars. The simplest naturally formed organic compounds reacted to form polymers and eventually compounds of greater and greater complexity.

The subject of life's origin is highly complex, but it is no longer the mystery it once was, in the early post-Darwinian period. In fact, there is no longer any fundamental difficulty in explaining, on the basis of physical and chemical laws, the origin of life from inanimate matter.

Darwin's Theory of Common Descent

After returning from his *Beagle* voyage in the 1830s, Darwin had concluded that the three species of mockingbirds on the Galápagos Islands must have been derived from a single mockingbird species on the South American mainland. Thus, a species could produce multiple descendant species. It was only a small step from this discovery to the postulate that all mockingbirds were derived from a common ancestor, and likewise all songbirds, all birds, all vertebrates, all animals, and finally, all life. Every group of organisms had descended from one common ancestral species. What was novel in Darwin's theory was that he proposed a branching phylogenetic tree, in contrast with the single linear ladder of the scala naturae that had been so widely supported in the eighteenth century.

Darwin's theory was persuasive because it supplied an explanation for numerous biological phenomena which up to that time had to be recorded as simply curious aspects of the world or as evidence for the planning of the Creator. First, Darwin's theory of common descent supplied the explanation for the findings of the comparative anatomists, particularly Cuvier and Owen, that organisms fall into well-defined groups that are constructed according to a common *bauplan* (or structural type or morphotype) and that permit the reconstruction of a definite archetype for each group. The theory of evolution through common descent also explained the origin of the Linnaean hierarchy, and it explained, most convincingly, the pattern of geographical distribution of the biota owing to the gradual spread of organisms onto all continents and their adaptive radiation in the newly settled areas.

Common descent has become the theoretical backbone of Darwinian evolutionary thinking since the publication of the *Origin,* not surprisingly so because it has such extraordinary explanatory powers. Indeed, the manifestations of common descent, as revealed by comparative anatomy, comparative embryology, systematics, and biogeography, were so convincing that evolution through common descent was accepted by the majority of biologists within the first decade of the publication of the *Origin.*

How far one could extend the common origin was at first controversial, even though Darwin himself suggested that "all our plants and animals [have descended] from some one form into which life was first breathed." Soon, indeed, protists were discovered that combined animal and plant characteristics, so much so that the classification of some of these intermediates is still debated. The capstone in the theory of common descent was provided in this century by the molecular biologists when they discovered that even bacteria, which have no nucleus, nevertheless have the same genetic code as protists, fungi, animals, and plants.

The theory of common descent had an enormously stimulating

influence on taxonomy (see Chapter 7). It suggested that one should try to find the nearest relative of every group of organisms, particularly isolated ones, and to reconstruct their common ancestor. This was more suggestive for animals than for plants, and certainly the construction of phylogenies was the favorite preoccupation of zoologists in the post-Darwinian period. In particular, it stimulated comparative researches in which every structure and organ was studied for the possibility that it was homologous with a corresponding structure in a related or possibly ancestral organism. A structure was considered homologous with that of another organism if both stemmed phylogenetically from a corresponding structure or characteristic of the putative immediate common ancestor. When the relationship of two groups was established by this method, as in the case of reptiles and birds, for example, investigators attempted to predict what the common ancestor would have looked like. It was an occasion for great rejoicing when such a "missing link" was found in the fossil record, as happened with *Archaeopteryx* in 1861, a fossil that was part bird, part reptile. Not that *Archaeopteryx* was necessarily a direct ancestor, but it indicates by what stages the transition might have occurred. ·

These studies were extended to the comparative study of embryos, and it was soon found, as emphasized particularly by Ernst Haeckel, that the course of individual (ontogenetic) development often went through stages similar to corresponding stages in an ancestral group. Hence, for instance, all terrestrial tetrapods during their ontogeny go through a gill arch stage, thus recapitulating (so to speak) the development of gills in their fish ancestors. A mild version of a theory of recapitulation has a good deal of validity, though it is not true that animals in their ontogeny recapitulate the adult stages of their ancestors (see Chapter 8).

In due time, it was possible to reconstruct a credible phylogenetic tree of the animals, while botanists, with the help of molecular evidence, are now on the way to doing the same for plants. Ultimately, this method was applied also to the prokaryotes, which were shown by Woese to consist of two major branches, the eubacteria and the archaebacteria. These findings have permitted the proposal of a new classification for all organisms (see Chapter 7).

THE ORIGIN OF HUMANS

Perhaps the most important consequence of the theory of common descent was the change in the position of man. For theologians and philosophers alike, man was a creature apart from the rest of life. Aristotle, Descartes, and Kant agreed on this, no matter how much they may have disagreed on other aspects of their philosophies. In the *Origin* Darwin confined himself to the cautiously cryptic remark, "Light will be thrown on the origin of Man and his history." But Haeckel (1866), Huxley (1863), and (in 1871) Darwin himself demonstrated conclusively that humans must have evolved from an apelike ancestor, thus putting our species into the phylogenetic tree of the animal kingdom. This effectively ended the anthropocentric tradition that had been maintained by the Bible and by most philosophers.

Darwin's Theory of the Multiplication of Species

According to the biological species concept, species are defined as aggregates of populations that are reproductively isolated from one another. This reproductive isolation is effected by certain species characteristics, including sterility barriers or behavioral incompatibilities, which are traditionally referred to as isolating mechanisms. They prevent the interbreeding of different species in areas where their ranges overlap. The problem of speciation is to explain how populations acquire such isolating mechanisms and how they can evolve gradually.[2] It is now almost universally agreed that the prevailing process of speciation is geographical, or allopatric, speciation—the genetic divergence of geographically isolated populations. It occurs in two forms: dichopatric speciation and peripatric speciation.

In dichopatric speciation, a previously continuous range of populations is disrupted by a newly arisen barrier (a mountain range, an arm of the sea, or a vegetational discontinuity). Either strictly by chance, as in the case of chromosomal incompatibilities, or by a change of function in behavior as a consequence of sexual selection (see below), or as an incidental byproduct of an ecological shift, the two separated populations will become genetically more and more different in time and, as a correlate of this difference, will acquire isolating mechanisms that will cause them to behave as different species when, later, they come again into contact with one another. It is now almost certain that most isolating mechanisms evolve prior to the time when the neospecies resume contact. The isolation may receive some additional fine-tuning after the secondary contact has been established, but the basic isolating factor originated prior to the contact.

In peripatric speciation, a founder population is established beyond the periphery of the previous species' range. Such a population, founded by a single inseminated female or by a few individuals, will contain only a small percentage of, and often an unusual combination of, the genes of the parent species. Simultaneously, it will be exposed to a new and frequently severe set of selection pressures owing to its changed physical and biotic environment. Such a founder population may undergo a drastic genetic modification and may speciate rapidly. Furthermore, such a founder population, owing to its narrow genetic base and drastic genetic restructuring, is in a particularly favorable position to undertake new evolutionary departures, including those that may lead to macroevolutionary developments.

In addition to these two forms of allopatric speciation, other scenarios have been proposed, and some of them may actually occur. The most likely of these processes is sympatric speciation, that is, the origin, owing to ecological specialization, of a new species within the cruising range of the individuals of the parental species. Highly improbable is so-called parapatric speciation, the development of a border between two species along an ecological escarpment within a species' range.

Darwin's Theory of Gradualism

Throughout his life, Darwin emphasized the gradual nature of evolutionary change. Not only was gradualness a necessary consequence of Lyell's uniformitarianism, but a sudden origin of new species would have seemed for Darwin too much of a concession to creationism. To be sure, at a given locality every species was sharply demarcated against other species, but when comparing geographically representative populations, varieties, or species, Darwin saw everywhere the evidence of gradualness.

Eventually, to us perhaps even more so than to Darwin, it became evident that evolution occurs in populations, and that sexual populations can change only gradually, never by a sudden saltation. There are some exceptions such as polyploidy, but they have never played a major role in macroevolution.

One of the most frequently raised objections to Darwin's gradualism was that it was unable to explain the origin of entirely new organs, structures, physiological capacities, and behavior patterns. For instance, how can a rudimentary wing be enlarged by natural selection before it can perform the functions of flying? Darwin proposed two processes by which such an evolutionary novelty can be acquired. One of them is what Severtsoff (1931) has called an intensification of function. Let us take as an example the origin of eyes. How could such a complex organ be created by natural selection? Eventually, it was shown that the earliest photoreceptor organs were simple light-sensitive spots on the epidermis, and that pigment, a lens-like thickening of the epidermis, and all the other accessory properties of eyes were gradually added in the course of evolution. Many of the intermediate stages are still in existence in various kinds of invertebrates. Such an intensification of function accounts for the various modification of the mammalian forelimbs in moles, whales, and bats, to mention just one other example.

However, another entirely different and much more dramatic way by which evolutionary novelties can be acquired is by a change in function of a structure. Here an existing structure, let us say the antennae of *Daphnia*, acquires the additional function of a swimming paddle and, under new selection pressure, become enlarged and modified. The feathers of birds presumably originated as modified reptilian scales serving for heat regulation but acquired a new function on the forelimbs and tails of birds in connection with flying.

During a succession of functions, a structure always passes through a stage when it can simultaneously perform both jobs. The antennae of *Daphnia* are a sense organ and a swimming paddle. Some of the most interesting examples of shift of function relate to behavior patterns, such as when the preening of feathers is incorporated into the courtship display of certain ducks. Many behavioral isolating mechanisms in animals probably originated through sexual selection in isolated populations and assumed their new function only after the species established contact with a related species.

MASS EXTINCTIONS

The discovery of mass extinctions was the second objection raised against Darwin's theory of gradualism. Prior to Darwin, the catastrophists, beginning with Cuvier, insisted that there had been a number of mass extinctions in which the then-ruling biota was decimated, if not totally exterminated, only to be replaced by a new biota. The fossil record suggested a considerable number of such drastic changes, as from the Permian to the Triassic or from the Cretaceous to the Tertiary. The major objective of Lyell's *Principles of Geology* was to refute catastrophism and to substantiate Hutton's thesis of gradual change in the history of the earth. Darwin's gradualism mirrored Lyell's view. Thus, it came as a rather unexpected development when mass extinctions were firmly documented exactly where the catastrophists had postulated them.

Mass extinctions are rare cataclysmic events superimposed on the normal Darwinian cycle of variation and selection leading to gradual change. Darwin was fully aware that the extinction of individual species and their replacement by new species is continuous throughout the history of life. But in addition to this background extinction, there were definite periods—which all along had served as demarcation lines between geological ages—when a large part of the biota became extinct simultaneously. The most drastic of these was at the end of the Permian, when more than 95 percent of all species died out completely.

The cause of mass extinctions is still debated today. The one at the end of the Cretaceous, which wiped out the dinosaurs, was almost surely the consequence of an asteroid impact and the climatic and other environmental changes it caused. This was first postulated by the physicist Walter Alvarez in 1980, but a great deal of supporting evidence has since been found. Indeed, the impact crater itself has been identified near the tip of the Yucatán peninsula. Endeavors to attribute the other mass extinctions to asteroid impacts have been unsuccessful. Most of them seem, rather, to be connected either with plate tectonic events that affected the size of shelf seas and the circulation of ocean currents or with other climatic changes. There is some regularity in the sequence of these extinctions, and some authors have postulated extraterrestrial causes, such as fluctuations in solar radiation—a plausible theory. However, most of the evidence for extraterrestrial explanations has not withstood critical analysis.

The species that have been lucky enough to survive a catastrophe that led to a mass extinction are like the members of a founder population. They have an entirely different biotic environment and can enter new evolutionary pathways. The most spectacular illustration of this possibility is provided at the beginning of the Tertiary, when an explosive radiation of the mammals—which had been on earth for more than one hundred million years before the extinction of the dinosaurs—occurred.

Darwin's Theory of Natural Selection

For a long time after Darwin's composite theory of the gradual evolution of species from a common ancestor had been widely accepted, a number of competing theories attempted to answer the question by what mechanism evolutionary change was effected. For some 80 years, the defenders of these theories were arguing with one another until during the evolutionary synthesis (see below) all the non-Darwinian explanations were so thoroughly refuted that Darwin's theory of natural selection was left as the only serious contender.

COMPETING THEORIES OF EVOLUTIONARY CHANGE

The three major non- or anti-Darwinian theories were saltationism, teleological theories, and Lamarckian theories.

Saltationism—a consequence of the typological thinking prevailing in the pre-Darwinian period—was supported by T. H. Huxley and Kölliker among Darwin's contemporaries, by the Mendelians (Bateson, De Vries, Johannsen), and by a few others (Goldschmidt, Willis,

Schindewolf) right into the period of the evolutionary synthesis. It was finally abandoned when population thinking was more widely adopted and when virtually no evidence for such a process of speciation could be found. A saltational origin of new species occurs in sexually reproducing organisms only through polyploidy and some other forms of chromosomal restructuring, and these are relatively rare forms of speciation.

Teleological theories claim that there is an intrinsic principle in nature which leads all evolutionary lineages to ever greater perfection. The so-called orthogenetic theories, such as Berg's nomogenesis, Osborn's aristogenesis, and Teilhard de Chardin's omega principle, are examples of teleological theories. They eventually lost all adherents when the haphazardness of evolutionary change (including many reversals) was demonstrated and when no mechanism could be found that could effect consistent progressive changes.

According to Lamarckian and neo-Lamarkian theories, organisms are slowly transformed during evolution through an inheritance of acquired characteristics. These new characteristics were believed to be due to the effects of use and disuse or, more directly, induced by forces of the environment. Since Lamarckism explained gradual evolution much better than the saltationism of the Mendelians, it was reasonably popular prior to the evolutionary synthesis. Indeed, right up to the 1930s there were probably more Lamarckians than Darwinians.

Lamarckian theories lost favor when the geneticists demonstrated that inheritance of acquired characters ("soft inheritance") cannot occur, because newly acquired characteristics of the phenotype cannot be transmitted to the next generation. The final demise of the theory of soft inheritance in the twentieth century was due to the finding of molecular biologists that the information contained in the proteins (the phenotype) cannot be transmitted to the nucleic acids (the genotype). This so-called central dogma of molecular biology deprived Lamarckism of the last remnants of credibility. There is a possibility that some microorganisms (perhaps up to the protists) have the ability to mutate in response to external conditions, but even if this should be confirmed, it would never be true for complex organisms, where the DNA of the genotype is too far removed from the phenotype.

NATURAL SELECTION

Darwinian natural selection is today almost universally accepted by biologists as the mechanism responsible for evolutionary change. It is best visualized as a two-step process: variation and selection proper.

The first step is the production of massive genetic variation in every generation owing to genetic recombination, gene flow, chance factors, and mutation. Variation, clearly, was the weakest point in Darwin's thinking. In spite of a great deal of study and hypothesizing, he never understood what the source of the variation was. He clearly had some erroneous ideas about the nature of variation, errors that were subsequently corrected by Weismann and by post-1900 genetics. We now know that genetic variation is "hard," not "soft," as Darwin thought. We also know that Mendelian inheritance is particulate—that the genetic contributions of the two parents do not blend when the egg is fertilized but remain discrete and constant. Finally, we have known since 1944 that the genetic material (composed of nucleic acids) is not directly converted into the phenotype but is merely the genetic information (the "blueprint" or program) which is translated into the proteins and other molecules of the phenotype.

The production of variation turned out to be a complex process. Nucleic acids can mutate (by changes in the base-pair composition) and do so copiously. Furthermore, during the formation of the gametes (meiosis) in sexually reproducing organisms, a process takes place by which the parental chromosomes are broken and reassembled. The resulting enormous amount of genetic recombination of the parental genotypes ensures that each offspring is unique. During this process of recombination, as well as in mutation, chance reigns supreme. There is a whole series of consecutive steps during meiosis where the assortment of genes is largely random and contributes a huge chance component to the process of natural selection.

The second step in natural selection is selection proper. This means the differential survival and reproduction of the newly formed individuals (zygotes). In every generation only a very small percentage of individuals in most species of organisms will survive, and certain individuals, owing to their genetic constitution, will under the prevailing circumstances have a greater probability of surviving and reproducing than others. Even in species where the two parents during their reproductive phase produce millions of offspring, as is the case for oysters and other marine organisms, on the average only two of them are needed to maintain the population at its steady-state frequency; and even if chance factors make a major contribution to the survival of these few progenitors of the next generation, there is no question that over time genetic properties make a major contribution to survival. In this manner, the adaptedness of the population is maintained from generation to generation, and the population is enabled to cope with environmental changes because certain genotypes among the vastly variable offspring are favored.

CHANCE OR NECESSITY?

From the Greeks to the nineteenth century there was a great controversy over the question whether changes in the world are due to chance or necessity. It was Darwin who found a brilliant solution to this old conundrum: they are due to both. In the production of variation chance dominates, while selection itself operates largely by necessity. Yet Darwin's choice of the term "selection" was unfortunate, because it suggests that there is some agent in nature who deliberately selects. Actually the "selected" individuals are simply those who remain alive after all the less well adapted or less fortunate individuals have been removed from the population. It has therefore been suggested that the term selection should be replaced by the phrase "nonrandom elimination." Even those who continue to use the word selection, which presumably will be the majority of evolutionists, should never forget that it really means nonrandom elimination, and that there is no selection force in nature. We use this term simply for the aggregate of adverse circumstances responsible for the elimination of some individuals. And, of course, such a "selection force" is a composite of environmental factors and phenotypic propensities. Darwinians take this for granted, but their opponents often attack a literal interpretation of these terms.

Only in recent years have evolutionists fully understood how drastically different Darwin's theory of evolution through natural selection was from earlier essentialistic or teleological theories. When Darwin

Darwin's explanatory model of evolution through natural selection

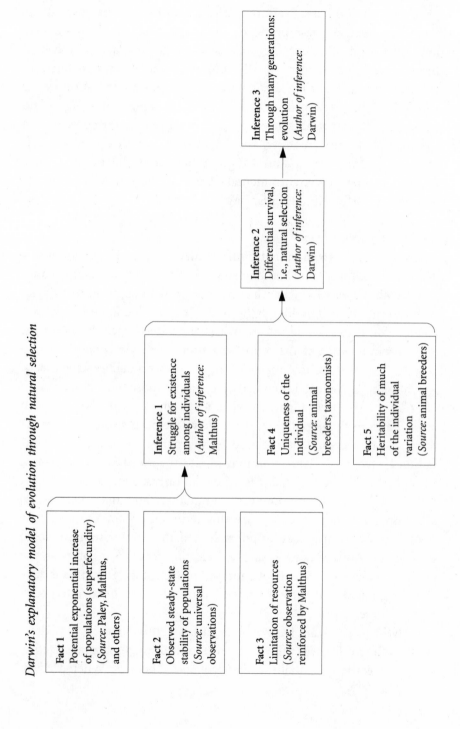

published the *Origin*, he had no proof for the existence of natural selection; he postulated it only by inference. Darwin's theory was based on five facts and three inferences (see diagram). The first three facts are the potentially exponential increase of populations, the observed steady-state stability of populations, and the limitation of resources. From this follows the inference that there must be competition (a struggle for existence) among individuals. Two further facts, the genetic uniqueness of every individual and the heritability of much of individual variation, lead to the second inference, differential survival (that is, natural selection), and to the third inference that a continuation of this process through many generations would result in evolution.

Darwin was delighted when Bates (1862) demonstrated the close resemblance and parallel geographic variation of palatable butterflies and their toxic, or at least unpalatable, models. This Batesian mimicry was the first clear proof of natural selection. Now, there are hundreds, if not thousands, of well-established proofs, including such well-known instances as insecticide resistance of agricultural pests, antibiotic resistance of bacteria, industrial melanism, the attenuation of the myxomatosis virus in Australia, the sickle-cell gene and other blood genes and malaria, to mention only a few spectacular cases.

The principle of natural selection is so logical and so obvious that today it can hardly be questioned at all. What can be, and indeed must be, tested in each individual case is to what extent natural selection has contributed to the characteristics of a particular component of the phenotype. For each characteristic, the questions that must be asked are: Was the evolutionary emergence of this characteristic favored by natural selection, and what was its survival value that has led to its being favored by natural selection? This is the so-called adaptational program.

SEXUAL SELECTION

Characteristics that favor survival include increased tolerance of adverse climatic conditions (cold, heat, drought), a better use of food resources, a greater competitive capacity, higher resistance to pathogens, and an increased ability to escape enemies. However, survival alone does not ensure an individual's genetic contribution to the next generation. From an evolutionary point of view, an individual may be more successful not by having superior survival attributes but merely by being more prolific reproductively. Darwin called the process whereby individuals were favored for their reproductive attributes "sexual selection."

He was particularly impressed by male secondary sexual characters such as the gorgeous plumes of male birds-of-paradise, the tail of the peacock, and the imposing antlers of stags. Today we know that the ability of females to choose their mate on the basis of these characteristics (a process called "female choice") is an important component of sexual selection, possibly more important than the ability of males to compete with rivals for access to females. Sexual and natural selection are not necessarily completely independent if, as seems to be the case, females are sometimes able to select males that will make a superior contribution to the survival value of their offspring.

Furthermore, there are other life-history phenomena such as sibling rivalry and parental investment that have more impact on reproductive success than on survival. Selection for reproductive success, thus, is apparently a larger category than the term sexual selection suggests. Much of what is studied by sociobiologists consists of aspects of selection for reproductive success.[3]

The Evolutionary Synthesis and After

For 80 years after Darwin published the *Origin*, controversy between Darwinians and non-Darwinians raged. One might have thought that the rediscovery of Mendel's genetic rules in 1900 would bring consensus, because of its potential for shedding light on variation, but it actually made the disagreement worse. The early Mendelians (Bateson, De Vries, and Johannsen) were incapable of population thinking and rejected gradual evolution and natural selection. The naturalists and biometricians who opposed them were no better. They accepted blending inheritance rather than the particulate inheritance that Mendel had demonstrated, and they waffled between natural selection and the inheritance of acquired characters. As late as the early 1930s several observers concluded that there was no hope for any consensus in the near future.

However, the groundwork for a consensus had been laid. Both the geneticists and the naturalists had separately greatly advanced our understanding of the origin of adaptedness and biodiversity, though neither camp knew much about the achievements of the other. Indeed, both had quite erroneous ideas about the other half of evolutionary biology. A bridge was needed, and in 1937 it was built through the publication of *Genetics and the Origin of Species* by Theodosius Dobzhansky. Dobzhansky was both a naturalist and a geneticist. In his youth in Russia he had been a beetle taxonomist and had become acquainted with the rich continental literature on species and speciation and had thoroughly absorbed population thinking. After 1927, when he had come to the United States to work in the lab of T. H. Morgan, he became well acquainted with the achievements and the thinking of the geneticists. As a result he was able in his book to do equal justice to both great branches of evolutionary biology: the maintenance (or improvement) of adaptedness by the turnover of genes in the gene pool, and the populational changes that lead to new biodiversity, particularly new species. The details in Dobzhansky's rough sketch were filled in by Mayr (species and speciation, 1942), Simpson (higher taxa, macroevolution, 1944), Huxley (1942), Rensch (1947), and Stebbins (plants, 1950). Simultaneously, there was a parallel synthesis in Germany, led by Timofeeff-Ressovsky, a student of Chetverikov.

Until the evolutionary synthesis, the study of macroevolution—evolution above the species level—was essentially in the hands of paleontologists who had no effective connection with either genetics or speciational studies. Hardly any paleontologists were strict Darwinians; most of them believed either in saltationism or in some form of finalistic autogenesis. They generally considered macroevolutionary processes and causations to be of a special kind, quite different from the populational phenomena studied by geneticists and students of speciation. This seemed to be confirmed by the prevalence of discontinuities among higher taxa—data seemingly in complete conflict with

Darwin's principle of gradualness. Everything in macroevolution seemed to be different from what one observed in microevolution.

Most students of macroevolution were still thinking in terms of transformational evolution, that is, a gradual change of evolutionary lineages toward ever greater specialization or adaptedness. However, they were confronted by Darwin's dilemma that the fossil record did not at all support this concept. On the contrary, long, continued, gradual changes of phyletic lineages were rare, if they existed at all. Instead, new species and higher types invariably turned up in the fossil record very suddenly, and most lineages became extinct sooner or later. To be sure, one could invoke the incompleteness of the fossil record, but since this seemed too much like sweeping a valid objection under the rug, many paleontologists adopted saltationism and were gratified when geneticists such as de Vries and Goldschmidt postulated evolution by macromutations ("hopeful monsters").

Simpson (1944:206) tried another solution, which he called quantum evolution, "the relatively rapid shift of a biotic population in disequilibrium to an equilibrium distinctly unlike an ancestral condition." He thought that this explained the well-known observation that "major transitions do take place at relatively great rates over short periods of time and in special circumstances." It is evident from his writings that Simpson had in mind a great acceleration of evolutionary change within a phyletic line. This type of solution was necessitated for him by his strictly phyletic species concept. Quantum evolution was criticized as being a return to saltationism, and Simpson more or less abandoned it in later years (1953).

EXPLAINING MACROEVOLUTION

With the evolutionary synthesis's refutation of saltationism, auto-[4] genetic theories, and soft inheritance, it became increasingly necessary to explain macroevolution as a populational phenomenon, that is, as a phenomenon that could be derived directly from events and processes occurring during microevolution.[4] This was particularly true in the case of the seeming saltations found in the fossil record. Paleontologists had neither the information nor the conceptual equipment to solve this problem.

In 1954 I proposed a solution: that the genetic restructuring takes place during a speciational process in founder populations and that some of the gaps in the fossil record are due to the fact that speciating founder populations, very much restricted in space and time, are most unlikely ever to be found in the fossil record. And yet, it is precisely peripherally isolated speciating populations that should be of most interest to students of major evolutionary changes.

It seemed to me that many puzzling phenomena, particularly those that concern the paleontologist, are elucidated by a consideration of these missing founder populations—including unequal (and particularly very rapid) evolutionary rates, breaks in evolutionary sequences and apparent saltations, and finally the origin of new "types." The genetic reorganization of peripherally isolated populations permits evolutionary changes that are many times more rapid than the changes within populations that are part of a continuous system. Here, then, is a mechanism that would permit the rapid emergence of macroevolutionary novelties without any conflict with the observed facts of genetics.[5]

This suggestion was taken up in 1971 by Eldredge and in 1972 by Eldredge and Gould, who gave the term "punctuated equilibria" to this kind of speciational evolution. They concluded, further, that if a new species originating by this progress was successful, it might then enter a phase of stasis and remain virtually unchanged for many millions of years until finally becoming extinct. Thus, macroevolution is not a form of transformational evolution but is as much a Darwinian type of variational evolution as is evolution within a species. There is a continuous production of new populations, most of which become extinct sooner or later. A certain percentage reach species level, mostly without acquiring any noteworthy evolutionary innovations, but also become extinct eventually. Only very rarely one of the new species acquires, during the period of genetic restructuring and the subsequent period of strong natural selection, a genotype that permits this neospecies to flourish, spread widely, and become a new component of the fossil record.

The paper by Eldredge and Gould had the effect that paleontologists finally became fully acquainted with the phenomenon of speciational evolution and began to understand why there are so many gaps in the fossil record. But perhaps the most important contribution which the theory of punctuated equilibria made is to call attention to the frequency of stasis. The endeavor of some geneticists to explain it as being due to normalizing selection is, of course, no explanation. Every population is continually subjected to normalizing selection. Yet, some of these populations and species evolve rapidly in spite of normalizing selection, while others remain phenotypically unchanged for many millions of years. One is forced to assume that such a stable phenotype is the product of a particularly well-balanced, internally cohesive genotype.

There are, indeed, many phenomena in the history of life which suggest the actual existence of such an internal cohesion. How else can one explain the virtual explosion of different structural types at the end of the Precambrian and the early Cambrian? Even in the utterly incomplete fossil record, one can distinguish at that time some 60 to 80 different morphotypes, compared with the 30 or so animal phyla now in existence. It seems as if the genotype of the new kingdom of animals was at first sufficiently flexible to produce, one might almost say experimentally, a high number of new types, some of which were not successful and became extinct, while the remaining ones, represented by the modern chordates, echinoderms, arthropods, and so on, became more and more inflexible. There has not been the production of a single major new bauplan since the early Paleozoic. It seems as if the existing ones had "congealed"—that is, had acquired such a firm internal cohesion that they could no longer experiment with the production of entirely new structural types.

Early in the history of genetics, it was realized that most genes are pleiotropic, that is, they have effects on several different aspects of the phenotype. Likewise, it was found that most components of the phenotype are polygenic characters, that is, affected by multiple genes. These interactions among genes are of decisive importance for the fitness of individuals and for the effects of selection, yet they are singularly difficult to analyze. Most population geneticists still restrict themselves to the study of additive gene phenomena and to the analysis of single-gene loci. This is understandable because the study of phenomena such as evolutionary stasis and the constancy of structural types are so refractory to genetic analysis. To acquire a better understanding of the cohesion of the genotype and its role in evolution is perhaps the most challenging problem of evolutionary biology.

Does Evolution Progress?

Most Darwinians have discerned a progressive element in the history of life on earth, as reflected in the progress from the prokaryotes, which dominated the living world for more than two billion years, to the eukaryotes, with their well-organized nucleus, chromosomes, and cytoplasmic organelles; from the single-cell eukaryotes (protists) to plants and animals with a strict division of labor among their highly specialized organ systems; within the animals, from ectotherms, which are at the mercy of climate, to the warm-blooded endotherms; and, within the endotherms, from types with a small brain and low social organization to those with a very large central nervous system, highly developed parental care, and the capacity for transmitting information from generation to generation.

Is it legitimate to call these changes in the history of life progress? This depends on one's concept and definition of progress. Yet, such change is virtually a necessity under the concept of natural selection because the combined forces of competition and natural selection leave little alternative but either extinction or evolutionary progression.

This change in the history of life is analogous to certain changes in industrial development. Why are modern motor cars so strikingly better than those of 75 years ago? Not because of any internal tendency of motor cars to get better but because manufacturers constantly experimented with various innovations, while competition through customer demand led to enormous selection pressure. Neither in the automobile industry nor in the world of life do we find any finalistic forces at work nor any mechanistic determinism. Evolutionary progress is simply the inevitable result of the simple Darwinian principle of variation and selection. It altogether lacks the ideological component one finds in the progressionism of the teleologists (like Spencer) and the orthogenesists.

It is curious how many people seem to have difficulty understanding a purely mechanistic path toward progress as represented by Darwinian evolution, on which developments are different in each phyletic lineage. Some lineages such as the prokaryotes have hardly changed at all for billions of years. Others have become highly specialized without showing any indications of being progressive, and still others, like most parasites and inhabitants of special niches, seem to have experienced a retrograde evolution. There simply is no indication in the history of life of any universal trend to, or capacity for, evolutionary progress. Where seeming progress is found, it is simply a byproduct of changes effected by natural selection.

WHY ORGANISMS ARE NOT PERFECT

If natural selection does not necessarily produce evolutionary progress, neither does it produce perfection, as Darwin pointed out. The limits to the effectiveness of natural selection are most clearly revealed by the universality of extinction: more than 99.9 percent of all evolutionary lines that once existed on earth have become extinct. Mass extinctions remind us forcefully that evolution is not, as conceived by transformational evolution, a steady approach to an ever higher perfection but an unpredictable process in which "the best" may be suddenly exterminated by a catastrophe and the evolutionary continuity taken over by phyletic lineages that prior to the catastrophe seemed to be without distinction or prospects.

Although, as Darwin pointed out, "natural selection is daily and hourly scrutinizing, throughout the world, every variation, even the slightest," there are, nevertheless, numerous limits or constraints on its power to bring about change.

First of all, the genetic variation needed to perfect a characteristic may not be forthcoming. Second, during evolution, the adoption of one among several possible solutions to a new environmental opportunity may greatly restrict the possibilities for subsequent evolution, as Cuvier already pointed out. For instance, when a selective advantage for a skeleton developed among the ancestors of the vertebrates and the arthropods, the ancestors of the arthropods had the prerequisites for developing an external skeleton, and those of the vertebrates had the prerequisites for acquiring an internal skeleton. The entire subsequent history of these two large groups of organisms was affected by the two different paths taken by their remote ancestors. The vertebrates were able to develop such huge creatures as dinosaurs, elephants, and whales, while a large crab is the largest type that the arthropods were able to achieve.

Another constraint on natural selection is developmental interaction. The different components of the phenotype are not independent of one another, and none of them responds to selection without interacting with the others. The whole developmental machinery is a single interacting system. This was realized by students of morphology as far back as Geoffroy St. Hilaire (1818), who expressed this in his *Loi de balancement.* Organisms are compromises among competing demands. How far a particular structure or organ can respond to the forces of selection depends, to a considerable extent, on the resistance offered by other structures, as well as other components of the genotype. Roux referred to the competitive developmental interactions as the struggle of parts in an organism.

The structure of the genotype itself imposes limits on the power of natural selection. The classical metaphor of the genotype was that of a beaded string on which the genes were lined up like pearls in a necklace. According to this view, each gene was more or less independent of the others. Not much is left of this previously accepted image. It is now known that there are different functional classes of genes, some charged to produce material, others to regulate it, and still others that are apparently not functioning at all. There are single coding genes, middle repetitive DNA, highly repetitive DNA, transposons, exons, and introns, and many other kinds of DNA. Discovering exactly how they all interact with one another, and particularly, what controls the epistatic interactions between different gene loci, is still a rather poorly understood area of genetics.

A further constraint on natural selection is the capacity for nongenetic modification. The more plastic the phenotype is (owing to developmental flexibility), the more this reduces the force of adverse selection pressures. Plants, and particularly microorganisms, have a far greater capacity for phenotypic modification (a wider reaction norm) than do animals. Natural selection is, of course, involved even in this phenomenon, since the capacity for nongenetic adaptation is under strict genetic control. When a population shifts to a new specialized environment, genes will be selected during the ensuing generations which reinforce and may, eventually, largely replace the capacity for nongenetic adaptation.

Finally, much of the differential survival and reproduction in a population is the result of chance, and this also limits the power of natural selection. Chance operates at every level of the process of reproduction, beginning with the crossing-over among parental chromosomes during meiosis to the survival of the newly formed zygotes. Furthermore, potentially favorable gene combinations are often destroyed by indiscriminate environmental forces such as storms, floods, earthquakes, or volcanic eruptions, without natural selection being given the opportunity to favor these genotypes. Yet over time, in the survival of those few individuals who become the progenitors of subsequent generations, relative fitness always plays a major role.

Current Controversies

The evolutionary synthesis thoroughly confirmed Darwin's basic principle that evolution is due to genetic variation and natural selection. However, within this basic Darwinian framework there is still room for considerable disagreement.

For many years a controversy has raged as to what the "unit of selection" was. Those who first adopted the term "unit" never explained exactly why they did so; in physics and technology a unit specifies the quantification of forces. The term unit in selection theory has a very different meaning, and what is worse, it is used in most discussions for two very different phenomena. They are "selection of," meaning the gene, individual, or group that is the target of selection, and "selection for," referring to a special characteristic or property, such as thick fur, favored by selection (sometimes conferred by a single gene).

The term unit is inappropriate when the question is whether the gene, the individual, the species, or what not is the target of selection. For most purposes of the evolutionist who wants to indicate what is selected (gene, individual, species), the term "target" is unquestionably more appropriate. However, even this word does not include every- thing the term "unit of selection" is supposed to cover. Here is clearly an area in need of greater conceptual clarification and terminological precision.

Most geneticists, for the convenience of their calculations, have considered the gene the target of selection and have tended to view evolution as a change in gene frequencies. The naturalists have con- tinued to firmly insist that it is the individual as a whole that is the principal target of selection and that evolution should be considered as the twin processes of adaptive change and the origin of diversity. Since no gene is ever directly exposed to selection, but only in the context of its entire genotype, and since a gene may have different selective values in different genotypes, it would seem highly unsuitable as the target of selection.

The proponents of "neutral evolution" are among the strongest supporters of the gene as the target of selection. The study of allozymes by the method of electrophoresis in the 1960s revealed a far greater amount of genetic variability than anyone had previously suspected. Kimura, as well as King and Jukes, concluded from this fact and from other observations that much of the genetic variation must be "neu- tral." This means that the newly mutated allele does not change the selective value of the phenotype. There is considerable argument whether the frequency of neutral mutations is really as great as claimed by Kimura (1983). What is far more controversial, however, is the evolutionary significance of neutral allele replacement. The neutralists, because they consider the gene to be the target of selection, regard neutral evolution as a highly important phenomenon. The naturalists, however, insist that with the individual as a whole being the target of selection, evolution takes place only if the properties of the individual change. A replacement of neutral genes is considered by them merely evolutionary "noise" and irrelevant for phenotypic evolution. If an individual is favored by selection owing to the overall qualities of its genotype, it is irrelevant how many neutral genes it may carry along as "hitchhikers." For the naturalists, such so-called neutral evolution is not in any conflict with the Darwinian theory.

GROUP SELECTION

There has been considerable uncertainty in the recent literature whether, in addition to individuals, also entire populations and even species could be the target of selection. Much of this controversy took place under the heading "group selection." The question was whether a group as a whole could be the target of selection, independent of the selective values of the individuals of which it is composed. To adequately approach the question, one should make a distinction between soft and hard group selection.

Soft group selection occurs whenever a particular group has more (or less) reproductive success than other groups simply because this success is due entirely to the mean selective value of the individuals of which the group is composed. Since every individual in sexually reproducing species belongs to a reproductive community, it follows that every case of individual selection is also a case of soft group selection, and nothing is gained by preferring the term soft group selection to the clearer traditional term individual selection.

Hard group selection occurs when the group as a whole has certain adaptive group characteristics that are not the simple sum of the fitness contributions of the individual members. The selective advantage of such a group is greater than the arithmetic mean of the selective values of the individual members. Such hard group selection occurs only when there is social facilitation among the members of the group or, in the case of the human species, the group has a culture which adds or detracts from the mean fitness value of the members of the cultural group. In animals, such hard group selection is found when there is a division of labor or mutual cooperation among the members. For instance, a group that has sentinels which warn of predators may gain in safety; another group may increase its survival by cooperating in the search for food, finding safe places for roosting, or through other cooperative aspects of communal living. In such cases of hard group selection, the application of the term group selection is justified.

Controversy has also swirled around the status of so-called species selection. Very often, the appearance of a new species seems to be instrumental in the extinction of another species. The success of certain new species has been designated as species selection. There is some justification in this term since, from the point of view of success, the new species seems to have a survival capacity superior to that of the old one. However, since the mechanism by which the species replacement is effected is individual selection, it might be less confusing to avoid the dual use of the term selection. For this reason, I prefer the terms species turnover or species replacement. Regardless of which term one uses, there is no doubt that this is a conspicuous aspect of evolutionary change and is of particular importance in macroevolution. It takes place strictly under Darwinian principles.

SOCIOBIOLOGY

The publication in 1975 of E. O. Wilson's book *Sociobiology: The New Synthesis* created a heated controversy around the question of what role evolution played in social behavior. Wilson, one of the foremost students of the behavior of social insects, came to the conclusion that social behavior merited far more attention than it had so far received, and that, indeed, its study deserved to be the subject matter of a special biological discipline which he baptized sociobiology. He defined it "as the systematic study of the biological basis of all social behavior." Ruse, in his book *Sociobiology: Sense or Nonsense* (1979a), defined it as "the study of the biological nature and foundations of animal behavior, more precisely, animal social behavior."

Wilson's work became highly controversial for two reasons. First, he included human behavior in his treatment and frequently applied the findings he had made in animals to the human species. The other reason was that both he and Ruse used the phrase "biological basis" in a somewhat equivocal manner. For Wilson, a biological basis for behavior meant that a genetic disposition makes a contribution to the behavior phenotype. For his politically motivated opponents, however, a biological basis meant "genetically determined." Of course we humans would be mere genetic automata if all of our actions were strictly and exclusively controlled by genes. Everyone (including Wilson) knows that this is not the case, and yet we also know, particularly from twin and adoption studies, that our genetic heritage makes a remarkably large contribution to our attitudes, qualities, and propensities. The modern biologist knows far too much to want to revive the old polarized nature–nurture controversy, because he knows that almost all human traits are influenced by the interaction of inheritance with the cultural environment. The most important point made by Wilson is that, in many respects, the same problems are encountered in human behavior studies as in animal ones. Likewise, many of the answers that have been found to be correct for animal behavior are applicable also in the study of human behavior.

According to the definitions of sociobiology offered by Wilson and Ruse, one would think that the field encompasses all social actions and interactions found in animals. This would include, for instance, all social migrations, such as those of African ungulates, of migratory social birds, the spawning migrations of horseshoe crabs and other invertebrates and vertebrates (such as gray whales). These and many other social phenomena, however, are not dealt with by Wilson and Ruse. Rather the subject matter of sociobiology is, according to Ruse, aggression, sex and sexual selection, parental investment, female reproductive strategies, altruism, kin selection, parental manipulation, and reciprocal altruism.

Most of these relate to the interaction between two individuals and deal either directly or indirectly with reproductive success. They all represent activities ultimately enhancing or reducing reproduction success, and they are, broadly speaking, related to sexual selection.

Sociobiology, thus circumscribed, is obviously a very special segment of the whole field of social behavior, and as such it raises all sorts of questions. Which kind of interactions between two individuals qualify as social behavior? When, if ever, is competition for resources social behavior? If sibling rivalry, which is competition for resources, qualifies as social behavior, when is competition not social behavior?

Most of the attacks against sociobiology were directed against its application to man. In Ruse's treatment two-thirds as many pages were devoted to human social behavior as to that of animals. This is the major reason for the controversial status of sociobiology and explains why most active people who work on the problems listed by Wilson and Ruse under sociobiology do not use that term for the type of work they do; they do not call themselves sociobiologists.

MOLECULAR BIOLOGY

Finally, in recent years much energy has been generated over the question to what extent the new findings of molecular biology require a revision of current evolutionary theory. It is sometimes stated that the findings of molecular biology necessitate a modification of the Darwinian theory. This is not the case. The findings of molecular biology relevant to evolution deal with the nature, origin, and amount of genetic variation. Some of the findings, such as the existence of transposons, are surprising, but all of the variation produced by these new molecular discoveries is ultimately exposed to natural selection, and is thus part of the Darwinian process.

The molecular discoveries of the greatest evolutionary importance are the following: (1) The genetic program (DNA) does not by itself supply the building material of a new organism, but is only the blueprint (information) for making the phenotype. (2) The pathway from nucleic acids to proteins is a one-way street. Information that proteins may have acquired is not translated back into nucleic acids; there is no "soft inheritance." (3) Not only the genetic code but, in fact, most of the basic molecular mechanisms are the same in all organisms, from the most primitive prokaryotes up.

N3

MULTIPLE CAUSES, MULTIPLE SOLUTIONS

Many of the controversies in biology that have arisen since Darwin's day owe their resolution to two important modifications in the way evolutionists think. The first is recognizing the importance of multiple simultaneous causes. Again and again an evolutionary problem seemed contradictory when only the proximate or the evolutionary cause was considered, while in reality the outcome was the result of the simultaneous occurrence of both proximate and ultimate causes. Similarly, other controversies erupted and then were resolved only when it was realized that both chance phenomena and selection occur simultaneously, or that geography and the genetic changes of populations together affect the speciation process.

In addition to having multiple causations, almost all evolutionary challenges have multiple solutions, and the recognition of this possibility has resolved many disputes. During speciation, for instance, pre-mating isolating mechanisms originate first in some groups of organisms, while in other groups post-mating mechanisms originate first. Sometimes geographic races are phenotypically as distinct as good species, yet not at all reproductively isolated; on the other hand, phenotypically indistinguishable species (sibling species) may be fully isolated reproductively. Polyploidy or asexual reproduction is important in some groups of organisms but totally absent in others. Chromosomal reconstruction seems to be an important component of speciation in some groups of organisms, but does not occur in others. Some groups speciate profusely; in others speciation seems to be a rare event. Gene flow is rampant in some species and drastically reduced in others. One phyletic lineage may evolve very rapidly, while other closely related ones may experience complete stasis for many millions of years.

In short, there are multiple possible solutions for many evolutionary challenges, even though all of them are compatible with the Darwinian paradigm. The lesson one must learn from this pluralism is that in evolutionary biology sweeping generalizations are rarely correct. Even when something occurs "usually," this does not mean that it must occur always.

CHAPTER TEN

What Questions Does Ecology Ask?

E cology, among all biological disciplines, is the most heterogeneous and most comprehensive. Almost everyone would agree that it deals with the interactions between organisms and their living as well as nonliving environment, but this definition permits an enormous range of possible inclusion. What, then, is the proper subject matter of ecology?[1]

The term "ecology" was coined by Haeckel in 1866 for the "household of nature." In 1869 he proposed a more elaborate definition: "By ecology we mean the body of knowledge concerning the economy of nature—the investigation of the total relations of the animal both to its inorganic and to its organic environment, including, above all, its friendly and inimical relations with those animals and plants with which it comes directly or indirectly into contact—in a word, ecology is the study of all those complex interrelations referred by Darwin as the conditions of the struggle for existence."

Despite this baptism by Haeckel, ecology did not become a truly active field until after about 1920; the founding of ecological societies and of professional journals devoted to ecology is of even more recent date. But to look at ecology from another point of view, it is nothing but "self-conscious natural history," as one ecologist has called it, and an interest in natural history goes back to primitive man.[2] Anything of concern to a naturalist—life history, reproductive behavior, parasitism, enemy thwarting, and so on—is automatically of equal concern to an ecologist.

A Brief History of Ecology

From Aristotle to Linnaeus and Buffon, natural history was largely descriptive, but not entirely so. In addition to their observations, naturalists also made comparisons and suggested explanatory theories that usually reflected the prevailing Zeitgeist. The great era of natural history was the eighteenth and first half of the nineteenth centuries, and the dominant ideology was natural theology.

According to this worldview, all in nature is in harmony, for God would not have permitted anything else. The struggle for existence was benign, programmed to maintain the balance of nature. Even though each set of parents produced an excessive number of offspring, they were reduced to the number required to sustain a steady-state population. The factors responsible for this reduction of numbers in every generation consisted of climatic causes, predation, diseases, failure to reproduce successfully, and so on. Nature operated, for the natural theologian, like a well-programmed machine. Ultimately everything could be attributed to the benevolence of the Creator. This worldview is well reflected in the writings of Linnaeus, William Paley, and William Kirby.

Gilbert White, the vicar of Selborne, is the eighteenth-century naturalist perhaps best known in the English-speaking world, but natural history flourished on the continent as well.[3] However, with the demise of natural theology in the middle of the nineteenth century and, more generally, with the steady strengthening of scientism, a largely descriptive natural history was no longer adequate. Natural history had to become explanatory. It continued to do what natural history had always done—observe and describe—but by applying other scientific methods to the observations (comparison, experiment, conjectures, testing of explanatory theories), it became ecology.

There were two major influences on the subsequent development of ecology: physicalism and evolution. The high prestige of physics as an explanatory science led to endeavors to reduce ecological phenomena to purely physical factors. This began with Alexander von Humboldt's ecological plant geography (1805) in which the overwhelming importance of temperature was stressed as a factor controlling both altitudinal and latitudinal composition of the vegetation (see below). His pioneering work was expanded by C. Hart Merriam (1894) in his endeavor to explain the vegetational zones on Mt. San Francisco in northern Arizona as a result of temperature. European plant geographers likewise stressed the importance of physical factors, particularly temperature and moisture.

The second major influence on ecology was the publication of Darwin's *Origin of Species.* Darwin completely refuted natural theology and explained the phenomena of nature by such concepts as competition, niche exclusion, predation, fecundity, adaptation, coevolution, and so on. He simultaneously rejected teleology, acknowledging the haphazardness of the fate of populations and species. Nature, as seen by Darwin and modern ecologists, is something entirely different from the God-controlled world of the natural theologian.

After Darwin, all physiological and behavioral adaptations of organisms—for their specialized mode of life or for the specialized environments in which they live—rightly came to be considered the concern of ecology. Some of the basic questions ecologists began to ask were: Why are there so many species? How do these species divide the resources of the environment among themselves? Why are most environments relatively stable most of the time? Are the well-being and population density of a species controlled more by physical factors or by biotic factors—the other species with which they live? What physiological, behavioral, and morphological properties permit a species to cope with its environment?

ECOLOGY TODAY

Modern ecology and its attendant controversies can be subdivided into three categories: the ecology of the individual, the ecology of the species (autecology and population biology), and the ecology of communities (synecology and ecosystem ecology). Traditionally zoologists have concentrated on autecological problems, botanists on synecological ones. Harper (1977) was one of the first botanists, if not *the* first, to study in plants the same kind of autecological problems that had concerned zoologists. But even now plant ecology is, on the whole, still a rather different field from animal ecology. And an ecology of fungi and prokaryotes hardly exists, at least under that name.[4]

The Ecology of the Individual

In the second half of the nineteenth century, as an extension of the activities of the naturalists, ecologists investigated the precise environmental requirements of individuals in a particular species: their climatic tolerance, life cycle, resources needed, and survival-controlling factors (enemies, competitors, diseases). They studied the adaptations which a given individual of a species must have to live successfully in its species-specific environment. These include hibernation, migration, nocturnal activity, and a host of other physiological mechanisms and behaviors that allow organisms to survive and reproduce under sometimes extreme conditions, ranging from the Arctic to deserts.[5]

From the standpoint of the ecology of the individual, the major role of the environment is to exercise continuously stabilizing selection, which weeds out all the individuals that have transgressed the permissible variation around the optimum. This is exactly what a Darwinian would expect. It is the environment, the biotic as well as the physical one, that plays a major role in natural selection. Every structure of an organism, each of its physiological properties, all of its behavior, and indeed almost any component of its phenotype and genotype has evolved for an optimal relation of the organism to its environment.

The Ecology of Species

After the ecology of the individual, the next development was the ecology of species, also called population biology. The local population—the population that is in contact with the populations of other species—is the object of special interest in this branch of ecology. What the ecologist, as population biologist, studies is the density of a population (number of individuals per unit area), the rate of increase (or decrease) of such a population under varying conditions, and, when dealing with the populations of a single species, all those parameters that control the size of a population such as birth rate, life expectancy, mortality, and so on.

This field can be traced back to a school of mathematical demographers interested in the growth of populations and the factors controlling it. Names associated with this movement are R. Pearl, V. Volterra, and A. J. Lotka.[6] Far more important for the practicing ecologist was the publication in 1927 of Charles Elton's *Animal Ecology,* "the sociology and economics of animals." From this date on, population biology was clearly recognized as a distinct subdiscipline of ecology.[7]

The population concept adopted by most mathematical population ecologists was basically typological, in that it neglected the genetic variation among the individuals of a population. Their "populations" were not populations in any genetic or evolutionary sense but were what mathematicians refer to as sets. The crucial aspect of the population concept to have emerged in evolutionary biology, by contrast, is the genetic uniqueness of the composing individuals. This kind of "population thinking" is in sharp contrast with the typological thinking of essentialism. In ecology the genetic uniqueness of the individuals of a population is usually ignored.

NICHE

A crucial characteristic of a species is that each one occupies a particular subdivision of the environment, which supplies all of its needs. The ecologist calls this the niche of the species. In the classical concept of the niche, developed by Joseph Grinnell, nature was visualized to consist of numerous niches, each suitable for a particular species. Charles Elton had a similar idea: the niche is a property of the environment.

Evelyn Hutchinson introduced a different concept of the niche. Although he defined it as a multidimensional resource space, his school, if I understand their writings, considered the niche more or less a property of the species. If a given species was absent in an area, it meant that its niche was also absent. But any naturalist studying a particular locality may discover insufficiently used resources or otherwise seemingly empty niches. This is well illustrated by the complete absence of woodpeckers in the forests of New Guinea, forests that in general structure and botanical composition are very similar to the forests of Borneo and Sumatra, where, respectively, 28 and 29 species of woodpeckers occur. Furthermore, the typical woodpecker niche does not seem to be filled in the New Guinea area by any other kind of bird. The same availability of unfilled niches is demonstrated by cases where an invading species seems to have little or no impact on the population size of the previously existing members of the community.

When one of a species' requirements is not adequately met—for instance, a chemical is missing from the soil, or heat is excessive—this "limiting resource" or "limiting factor" may prevent the existence of the species at that locality. The borders of a species' range (when not determined by geographic barriers) are usually controlled by such limiting factors as temperature, rainfall, soil chemistry, and the presence of predators. On continents, as Darwin well knew, species borders often seem to be due to competition with another species.

COMPETITION

When several individuals of the same species or of several different species depend on the same limited resource, a situation may arise that is referred to as competition. The existence of competition has been long known to naturalists; its effects were described by Darwin in considerable detail. Competition among individuals of the same species (intraspecies competition), one of the major mechanisms of natural selection, is the concern of evolutionary biology. Competition among the individuals of different species (interspecies competition) is a major concern of ecology. It is one of the factors controlling the size of competing populations, and in extreme cases it may lead to the extinction of one of the competing species. This was described by Darwin in the *Origin* for indigenous New Zealand species of animals and plants which died out when European competitors were introduced.

No serious competition exists when the major needed resource is in superabundant supply, as in most cases of the coexistence of herbivores. Furthermore, most species do not depend entirely on a single resource, and if the major one becomes scarce they shift to alternative resources and, in the case of competing species, usually to different ones. Competition is usually most severe among close relatives with similar demands on the environment, but it may also occur among totally unrelated forms that compete for the same resource, such as seed-eating rodents and ants. The effects of such competition are graphically demonstrated when entire faunas or floras come into competition, as happened at the end of the Pliocene when North and South America were joined at the Isthmus of Panama. It resulted in the extermination of a large fraction of the South American mammal fauna, apparently unable to withstand the competition from invading North American species, although added predation was also an important factor.

To what extent competition determines the composition of a community and the density of particular species has been the source of considerable controversy. The problem is that competition ordinarily cannot be observed directly but must be inferred from the spread or increase of one species and the concurrent reduction or disappearance of another species. The Russian biologist Gause performed numerous two-species experiments in the laboratory, in which one of the species became extinct when only a homogeneous resource was available. On the basis of these experiments and of field observations, the so-called law of competitive exclusion was formulated, according to which no two species can occupy the same niche. Numerous seeming exceptions to this "law" have since been found, but they can usually be explained as revealing that the two species, even though competing for a major joint resource, did not really occupy exactly the same niche.

Competition among species is of considerable evolutionary importance. It exerts a centrifugal selection pressure on coexisting species, resulting in morphological divergence among sympatric species as well as in a tendency to expand their niches into nonoverlapping areas. Darwin referred to this as the principle of divergence. Where the competition leads to the extinction of one of the species, it has been referred to as "species selection." However, species replacement or species turnover may describe the situation better, because selection pressures are brought to bear on the individuals of the competing species, even though the well-being and existence of entire species are affected. "Species selection" is actually a result of individual selection.

Competition may occur for any needed resource. In the case of animals it is usually food; in the case of forest plants it may be light; in the case of substrate inhabitants it may be space, as in many shallow-water benthic marine organisms. Indeed, it may be for any of the factors, physical as well as biotic, that are essential for organisms. Competition is usually the more severe the denser the population. Together with predation, it is the most important density-dependent factor in regulating population growth.

REPRODUCTIVE STRATEGIES AND POPULATION DENSITY

Population biologists have discovered that most species can be placed in one of two classes with respect to their population size and reproductive strategy. The first has strongly variable population size, often subject to catastrophes, with intraspecific competition weak. Such species tend to have high fertility; that is, they have adopted an r-selection strategy. Other species have a nearly constant population size from year to year, close to carrying capacity, and are subject to strong intra- and interspecific competition. They tend to have a longer life span and are selected for slow development, delayed reproduction, and single broodedness. This is referred to as the K-selection strategy.

Even taking these differences in reproductive strategy into account, the fertility of every species is so high that if all the offspring of a pair were to again reproduce, population size in due time would approach infinity. It has been known since ancient times, however, that only a fraction of the offspring of one generation survives to produce the next. Among the factors responsible for this reduction in every generation are competition for limited resources, climatic variations, predation, diseases, and failure to reproduce successfully. The result is that the populations of most species, in spite of variation and fluctuations and the continuing death of individuals, achieve a steady state. How this equilibrium is achieved has been the source of numerous controversies in the ecological literature.

Ecologists early realized (with evidence persuasively presented by

David Lack) that much mortality in natural populations is density-dependent. This means that as the density of a population increases, such adverse factors as predation, competition, disease, scarcity of food, and shortage of hiding places have greater impact and lead to increased mortality, thus slowing population growth. This discovery led to the view that populations have a self-regulating power,[8] through life-history constraints on population growth such as establishing territories, reducing clutch size in the case of birds, increasing dispersal in the case of some plants, and many others. However, this self-regulating power, in order to be able to operate, would require an assumption of group selection (see Chapter 8), a process which, after an initial period of popularity, was soon shown not to operate except in social species. Lack, G. C. Williams, and others showed that natural selection operating on individuals, together with kin selection (see Chapter 12), is sufficient to explain territoriality, low reproductive rate, dispersal, and all other known phenomena that were once attributed to self-regulation. The theory of self-regulation is no longer considered seriously.

Andrewartha and Birch claimed that climate could override all adverse density-related factors and could control population size in a density-independent way. Indeed, everybody knew that climatic factors such as severe winters, hot summers, droughts, and excessive precipitation could have a catastrophic impact on populations, particularly of insects and other invertebrates. A sophisticated statistical analysis of population changes uninfluenced by population density has shown that an effect of density is superimposed on the climate-induced population fluctuations. The size of populations evidently is controlled both by physical and biological factors.

PREDATORS, PREY, AND COEVOLUTION

While most species have remarkably similar population size from year to year, others are characterized by irregular or cyclic fluctuation of their populations. Elton (1924) showed that these fluctuations among small herbivores (mice, lemmings, hares) resulted in similar fluctuations in their predators, such as the Arctic fox. Small Arctic rodents usually have a three- to four-year cycle and so do their predators.

Hares, which are larger, often have nine- to ten-year cycles, as do their predators. We now know that the cycles of the herbivores produce the cycles of the predators, and not the reverse.

In response to predator pressure, prey usually acquire certain adaptive behaviors (for example, seeking and finding retreats) or they acquire better protection (heavier shells, for instance), distastefulness, and so on. Over time the predators, in turn, are selected to cope with these defenses. The result is a sort of escalating "arms race" between predator and prey. Most plants have developed a whole battery of defenses by various chemicals, particularly alkaloids, that make them unattractive to most herbivores, but there are usually a few taxa of herbivores that are able to cope with these chemical defenses.

When a plant evolves new chemicals in defense against herbivores and the herbivorous insects in turn evolve new detoxification mechanisms, we may speak of the "coevolution" of the interacting species. Coevolution can also take the form of mutualism, including symbiosis. A famous example of the latter is the yucca moth, which destroys some of the potential seeds of the yucca on behalf of the moth's developing larvae but in turn pollinates the flowers, ensuring both the well-being of her larvae and the production of a sufficient surviving supply of seeds.

There are cases where predators, particularly when newly introduced into an area, have had devastating effects on certain prey species. In some rare cases a predator may entirely wipe out a prey species, as when the cactus moth *(Cactoblastis)* virtually destroyed the introduced *Opuntia* populations in Queensland, Australia. Normally, some individuals survive, and the prey population recovers after the crash of the predator population. The manifold interactions between predators and prey are an active area of research in ecology. It is particularly important for the biological control of agricultural pests.

THE FOOD CHAIN AND THE PYRAMID OF NUMBERS

Elton pointed out that the members of a community actually form a food chain, with photosynthesizing plants being the first link, herbivores the next link, carnivores the third link, and decomposers (microbes and fungi) the last link. The photosynthetic plants are called

the producers, while the other members of the food chain are sometimes called the consumers. The carnivores, in turn, may consist of
different size classes, large carnivores often feeding not only on herbivores but also on small carnivores.

There is, on the average, an increase in size as one ascends the food
chain. Among the herbivores are myriads of insects (and their larvae),
while the carnivores are usually of larger size and much smaller in
number. However, examples such as the elephant and large ungulates
show that herbivores can also reach large size. Indeed, the largest
herbivores (elephants, large dinosaurs) are usually larger than the
largest coexisting carnivores.

The photosynthetic plants supply by far the greatest contribution
to the earth's biomass, while the herbivores contribute far less, and
the carnivores even less. The number of carnivores is quite small
compared with the herbivores they consume, and this leads to a
"pyramid of numbers," which reflects the fact that organisms at the
top of the food chain are comparatively sparse. A cat feeding on mice
or a whale feeding on millions of krill (Euphausia) illustrates this
reduction in numbers at higher levels in the food pyramid.

LIFE HISTORIES AND TAXONOMIC RESEARCH

All comparative studies on rarity of species, the size of a species' range,
predator–prey interactions, and the many other areas of research in
population biology depend upon knowledge of existing species taxa
and their life history. Most traditional naturalists, particularly botanists
and students of insects and aquatic organisms, were also taxonomists.
Indeed, their life-history studies of living organisms helped them
greatly in their taxonomic discrimination. Such dual competence became much rarer after the emancipation of ecology from natural
history, but all good taxonomists remained good naturalists.[9]

The study of the life histories of animals and plants is clearly a
long-standing concern of the ecologist. In the case of plants, the
classical division into annuals and perennials is based on a life-history
criterion, and the assignment of plants to the classes herbs, shrubs,
and trees likewise uses an ecological criterion. In the case of animals,
virtually all aspects of their life history—longevity, fertility, seden-

tariness, nature of the niche, seasonality, frequency of reproduction, mating systems, and so on—affect reproductive success and population size and hence are of interest to the population biologist.

Yet after several hundred years of dedicated work by taxonomists, we still have no reliable count of how many species exist, much less a working knowledge of the life history of each. If there are 10 million species of animals (a very conservative estimate), and if about 1.5 million species have been described, it would mean that about 15 percent are known. If, however, the number of species is 30 million (a distinctly legitimate estimate), then only 5 percent are known.

Moreover, the degree of knowledge of different groups is highly uneven. The number of species of birds is around 9,300, and most recent increases in this number were the result not of new discoveries but of elevating isolated populations to the rank of species. The number of new species of birds discovered in the last 10 years is less than one third of 1 percent of the total number. In other words, at least 99 percent of all existing species of birds have been discovered and described. By contrast, in many groups of insects, arachnids, and lower invertebrates the number of known species may be less than 10 percent of the number of actually existing species, and the same is true for fungi, protists, and prokaryotes. Studies on local species diversity are woefully insufficient for tropical biota and for special marine environments. This is one of the reasons why ecologists endorse so wholeheartedly increased support for taxonomic research.

The Ecology of Communities

A kind of ecology entirely different from that of individuals or populations began to evolve in the late nineteenth century, as ecology grew into an independent science from its roots in natural history and plant geography. The emphasis of this new "community ecology" or "synecology" was on the composition and structure of communities consisting of different species.[10]

Beginnings of this way of viewing nature can be found in the writings of Buffon. But the real founder of community ecology was Alexander von Humboldt in his analysis of vegetation types—types of vegetation created by similar climates regardless of the taxonomic relationship of the constituent species. Vegetation types include grasslands, temperate deciduous forest, tropical evergreen rainforest, tundra, and savannas; and since these were the most conspicuous examples of communities, the emphasis in synecology was on plant communities and was strongly geographical.

A rainforest, whether in the Australian region or in Amazonia, has a characteristic appearance, and so has a desert, regardless of the continent on which it occurs. Taxonomically, as Darwin remarked, the plants of a particular vegetation type, for instance deserts, on the different continents are not particularly related to one another but rather to plants of other vegetation types of the same continent. Plant ecologists from Humboldt on, but particularly beginning with the second half of the nineteenth century, have attempted to characterize these various vegetation types and their causes.

Eugene Warming's *Ecology of Plants* (1896) was the most successful product of this tradition, and Warming has been called the father of ecology. All members of his school were strongly physicalist in their explanations, with emphasis on the role of temperature, water, light, nitrogen, phosphorus, salt, and other chemicals in the distribution of vegetative types. But for Warming, in contrast to many of his predecessors, precipitation rather than temperature was the primary determinant. He had been led to this conclusion by his researches in the tropics. This type of ecology, strictly speaking, became known as the geographical ecology of plants.[11]

SUCCESSION AND CLIMAX

In the early twentieth century the American ecologist Frederic Clements was the first to point out that a succession of plant communities would develop after a disturbance such as a volcanic eruption, heavy flood, windstorm, or forest fire. An abandoned field, for instance, will be invaded successively by herbaceous plants, shrubs, and trees, eventually becoming a forest. Light-loving species are always among the first invaders, while shade-tolerant species appear later in the succession.

Clements and other early ecologists saw almost lawlike regularity in the order of succession, but that has not been substantiated. One of the most carefully documented studies of succession was the reestablishment of a biota on the island Krakatau, which was left totally barren by a volcanic eruption in 1883 (Thornton 1995). In this and other successions, a general trend can be recognized, but the details are usually unpredictable. One abandoned pasture in New England may be taken over by white pine and gray birch, another nearby pasture may first be invaded by junipers, bird cherries, and maples. Succession is influenced by many factors: the nature of the soil, exposure to sun and wind, regularity of precipitation, chance colonizations, and many other random processes. An early student of succession was the American naturalist and poet Henry David Thoreau (1993).

The final stage of a succession, called the climax by Clements and early ecologists, is likewise not predictable or of uniform composition. There is usually a good deal of turnover in species composition even in a mature community, and the nature of the climax is influenced by the same factors that influenced succession. Nevertheless, mature natural environments are usually in equilibrium and change relatively little through time unless the environment itself changes.

For Clements, the climax was a "superorganism," an organic entity.[12] Even some authors who accepted the climax concept rejected Clements's characterization of it as superorganism, and it is indeed a misleading metaphor. An ant colony may be legitimately called a superorganism because its communication system is so highly organized that the colony always works as a whole and appropriately according to the circumstances, but there is no evidence for such an interacting communicative network in a climax plant formation. Many authors prefer the term "association" to the term "community" in order to stress the looseness of the interaction.

Even less fortunate was the extension of this type of thinking to include animals as well as plants. This resulted in the "biome," a combination of coexisting flora and fauna. Though it is true that many animals are strictly associated with certain plants, it is misleading to speak of a "spruce–moose biome," for example, because there is no internal cohesion to their association as in an organism. The spruce community is not substantially affected by either the presence or absence of moose. Indeed, there are vast areas of spruce forest without moose. There has always been a somewhat mystical overtone to the description of plant communities as superorganisms.

The opposition to the Clementian concepts of plant ecology was initiated by Herbert Gleason (1926), soon joined by various other ecologists. Their major point was that the distribution of a given species was controlled by the niche requirements of that species and that therefore the vegetation types were a simple consequence of the ecologies of individual plant species.

ECOSYSTEM

With climax, biome, superorganism, and various other technical terms for the association of animals and plants at a given locality being criticized for one reason or another, the term ecosystem was more and more widely adopted. Proposed by the English plant ecologist A. G. Tansley (1935), the term refers to the whole system of associated organisms together with the physical factors of their environment.

R. Lindeman (1942) eventually emphasized the energy-transforming role of such a system. This was well described by one ecologist with the words, "An ecosystem involves the circulation, transformation, and accumulation of energy and matter through the medium of living things and their activities." Photosynthesis, decomposition, herbivory, predation, parasitism, and other symbiotic activities are among the principal biological processes responsible for the transport and storage of materials and energy. The ecologist, then, is "primarily concerned with the quantities of matter and energy that pass through a given ecosystem and with the rates at which they do so" (Evans 1956). It was the chief mission of the International Biological Program (IBP) to get such quantitative data.

Alas, it would seem that this physicalist approach was not a great improvement over its predecessors. Although the ecosystem concept was very popular in the 1950 and 60s, particularly owing to the enthusiasm of Eugene and Howard Odum, it is no longer the dominant paradigm. Gleason's arguments against climax and biome are largely valid also against ecosystems. Furthermore, the number of interactions is so great that they are difficult to analyze, even with the help of large computers.

Finally, most younger ecologists have found ecological problems involving behavior and life-history adaptations more attractive than measuring physical constants. Nevertheless, one still speaks of the ecosystem when referring to a local association of animals and plants, usually without paying much attention to the energy aspects. An ecosystem does not have the integrated unity one expects from a true system.

DIVERSITY

What factors, then, control the number of species that coexist at a given place? The most obvious generalization one can make is that the more demanding the environment is, the fewer species will make up the community. Hence, a severely stressed area, like a desert or the Arctic tundra, will have far fewer species than a subtropical or tropical forest. But this is not all. Evidently, historical factors, such as the origin of a biota as a result of the fusion of two previously separated biota, as well as the suitability of an area for speciation (such as having many potential geographic barriers), also have a strong influence. This may explain why a given area in Malaysia may have three times as many species of forest trees as an equivalent area in an Amazonian rainforest.

Two species may exclude each other at one locality and coexist peacefully at another one. Potential competitors may form so-called guilds, the specific composition of which may change from place to place. For instance, on the smaller islands east of New Guinea one may find a large, a medium-sized, and a small fruit-eating pigeon. However, which of a series of large pigeons, of medium-sized pigeons, and of small pigeons is represented on a given island is unpredictable and seemingly due to chance occurrences.

No matter how relatively stable a community may seem to be, it actually reflects a balance between extinction and new colonization. This was first clearly seen by students of island populations and was later formulated mathematically as the law of island biogeography. The smaller the island, the more rapid the turnover of species; conversely, the slower the turnover, the higher the percentage of endemic species. The longer a population can survive while being isolated on an island, the greater the probability that it will become a separate species.[13]

MacArthur claimed in 1955 that the more diverse a community was, the more stable it would be. May (1973) came to the opposite conclusion, and subsequent researches have not been able to arrive at a consensus. What is evident is that the composition of a community is the result of a highly complex interaction of historical, physical, and biotic factors and is in most cases predictable only very approximately. The factors that influence the composition, such as the physical characteristics of the environment and the presence of competitors and enemies, are usually apparent, but the relative importance of these factors may be strongly influenced by historical contingencies.

Paleoecology

As the study of fossil assemblages matured, paleontologists increasingly paid attention to the ecology of former biota. A number of ecological problems are particularly expressed in fossil biota, although the conclusions drawn in these researches are constrained by the problem of differential preservation. Soft body taxa fossilize only under rare conditions, but even those with shells and skeletons show considerable differences in preservation. Sometimes entire local communities are seemingly well preserved, for instance reef communities. The circumstances of deposit and preservation are investigated by the methods of taphonomy.

The most conspicuous area of interest in paleoecology is the extinction of entire major taxa. What caused the extinction of the trilobites, for instance, the dominant taxon of invertebrates in the Paleozoic? Or that of the ammonites, a nearly equally dominant group in the Mesozoic? If the termination of such a group coincides with one of the great periods of extinction in the earth's history, then one can ascribe it to the same cause as the general extinction. This is true, for instance, for the demise of the dinosaurs, which coincides with the end of the Cretaceous and, as is now more or less agreed upon, with the impact of the Alvarez asteroid in Yucatán. The extinction of the trilobites is often attributed to competition by the "functionally more efficient" mollusks, but this is largely a *post hoc, ergo propter hoc* inference.

Life on earth originated in water, and one of the greatest ecological revolutions was the conquest of land, first by plants and then by animals. But just as we had the replacement of trilobites and ammonites in the water, so we had major replacements on land. The upsurge of the mammals after the extinction of the dinosaurs is most frequently mentioned, but a far more drastic, although less complete, change occurred among land plants. The dominant vegetation consisting of tree ferns, horsetails, and gymnosperms was largely replaced during the Cretaceous by flowering plants (angiosperms). A credible scenario involving insect (instead of wind) pollination and seed dispersal by birds and mammals has been suggested by Regal (1977). It is interesting that in this scenario the change is credited not to physiological or climatic but to ecological factors.

Controversies in Ecology

Few of the major controversies in ecology, if any, have been decisively settled. What controls population density, competition, or predation? Are density-dependent factors more important or density-independent ones? Is there a terminal stage in a succession and how predictable is it? How rigid is the "law" of competitive exclusion? In all of these controversies there is now a dominant view but also a minority position. A change from one to the other can often take place very rapidly, as in the question whether or not the richest biota is the most stable one.

Pluralism seems to be the correct answer in many, if not most, controversies in ecology. Different kinds of organisms may obey different rules. Or different determining factors may prevail in aquatic environments and in terrestrial ones. Or the dominant factors may change with latitude. When two authors disagree with respect to the solution of an ecological problem, it is not necessarily true that one of the two must be wrong. Instead it may be a case of pluralism.

Other controversies in ecology, as in other areas of biology, result from a failure to recognize both proximate and evolutionary causations. Ecology differs from most other biological disciplines in that it does not squarely fit into the biology of either proximate or evolutionary causations. Moreover, parts of ecology, like evolutionary ecology, are dominated by an intricate synergism of proximate and ultimate causations. It is most important in the study of ecological phenomena to discriminate between the two causations, if one is to properly disentangle cause and effect.

Just as one must use population thinking to find the answer to evolutionary problems, so one must apply ecological thinking not just on behalf of conservation but with respect to all our dealings with the environment, including all economic issues in forestry, agriculture, fisheries, and so on. And one must always remember that it is a rare case that allows one to apply a simple recipe. Ecological interactions are often chain reactions, with the end result becoming apparent only after very detailed and sophisticated analysis. No one seems to have expected that the destruction of the seabird colonies on Novaja Zemlya by radioactive materials would lead to a collapse of the local fisheries. The introduction of exotic flora or fauna (such as the rabbit in Australia), whether deliberate or accidental, has often had unexpected catastrophic effects. Not all of this can be predicted or prevented by ecological research, but some of it clearly can, or at least it can be mitigated or reversed. Sometimes, a timely ecological analysis can prevent an act, such as the building of a dam, that would have disastrous consequences.

The appearance of civilized man has had an impact on nearly every previously natural plant community. Beginning with George Perkins Marsh and Aldo Leopold, naturalists have pointed out in how many different ways humans have caused sweeping changes in the natural vegetation. The deforestation of the mountains in the Mediterranean and today of the tropical rainforest, as well as overgrazing (particularly by goats) in many areas of the subtropics, has had drastic and very often disastrous effects on the natural landscape and its human inhabitants. This is what the conservation movement has called attention to, pointing out the measures (particularly population control) that are necessary to reduce further damage.

Like every other species, humans have their species-specific ecology. Four major areas concern the ecologist: (1) the dynamics and consequences of human population growth, (2) the use of resources, (3) the impact of human beings on their environment, and (4) the complex interactions between population growth and environmental impact. As ecologists and environmentalists have often pointed out, the problem of the future of humankind is ultimately an ecological problem.

CHAPTER ELEVEN

Where Do Humans Fit into Evolution?

In most primitive cultures, Greek philosophy, and, conspicuously, the Christian religion, humans were considered to be entirely apart from the rest of nature. Not until the eighteenth century did a few daring authors call attention to the similarity of Man and the apes; Linnaeus even included chimpanzees in the genus *Homo*. But perhaps the first person to postulate clearly the descent of humans from primates was the French naturalist Lamarck (1809). He even supplied a scheme to explain how humans had descended from the trees and acquired bipedalism, and how the form of the human face was altered through a change of diet.

But it was Darwin's theory of common descent which left no escape from the conclusion that humans indeed had descended from apelike ancestors; the comparative morphological evidence had become overwhelming. A few years later Huxley, Haeckel, and others firmly established the principle that there was nothing supernatural about the origin of human beings. No longer isolated from the rest of the living world, *Homo sapiens* and its evolutionary history had become secularized into a branch of science.

Slowly but inevitably a new biological discipline began to develop, human biology. It had multiple roots: physical anthropology, comparative anatomy, physiology, genetics, demography, cultural anthropology, psychology, and others. Its task was twofold: to show how humans are unique with respect to all other organisms, and yet to show how human characteristics evolved from those of our ancestors.

How could one resolve the seeming contradiction between the fact that humans were animals, and yet so fundamentally different from any other animals, even their closest relatives among the apes? The more carefully one studied humankind as well as the vast diversity of the world of life, the more one was impressed by the utter improbability of human beings. How could such an extraordinary creature have emerged from the animal kingdom?

Whenever human beings had been considered in the pre-Darwinian literature, for instance by Lamarck, their rise was always explained as the inevitable culmination of a trend toward ever-greater perfection; Man was the highest step in the *scala naturae.* But Darwin made this teleological interpretation unnecessary; his theory of natural selection explained mechanistically all the phenomena previously explicable only with the help of metaphysical concepts. Biological science was now clearly given a new task. It had to explain the gradual evolution of humans from their primate ancestors as the result of the ordinary evolutionary processes, particularly natural selection, that operated in the rest of the living world.

Another powerful ideology that was eventually eliminated from the study of human evolution after 1859 was essentialism. Darwin's new concept of population thinking, which stressed the uniqueness of each individual within a population, had to be applied also to human beings. Anthropologists were slow to do so, but whenever they did, the new guideline yielded magnificent results.

Still, much about the evolution of *Homo sapiens* is a riddle, even today. When and where did the hominid line branch off from the ape (pongid) line that led to the modern apes? After the separation from the ape line, through what stages did the hominid line pass before a truly human level was reached?

The Relationship of Humans to the Apes

In the first post-Darwinian tree constructions the branching point for the hominid line was postulated to have been very early. However, all efforts to find fossil hominids from 13 million to 25 million years ago (the Miocene epoch) were unsuccessful. For a while an Asiatic fossil primate, *Ramapithecus*, about 14 million years old, was thought to be closer to humans than to any of the apes, but eventually it was shown to belong to the line of the orangutan.

The finding of Neanderthal fossils in 1849 at Gibraltar marked the beginning of the study of early hominids. Over the next 40 years all fossil hominids found were either *Homo sapiens* or Neanderthal. Then in 1892 Dubois found an early hominid in Java, which he called *Pithecanthropus erectus*; its vicariant in China, Peking Man (named *Sinanthropus pekinensis*), was described in 1921. Both were later combined with African fossils into the species *Homo erectus* (see below).

But the true "missing link" was not found until 1924, when Dart described a fossil from south Africa which he considered intermediate between humans and apes. He named it *Australopithecus africanus.* From that time on, numerous australopithecine fossils have been found in east and south Africa. They are usually assigned to two branches, a gracile branch to which *Australopithecus africanus* belongs, and which eventually gave rise to the genus *Homo*, and a robust sidebranch represented in south Africa by *Australopithecus robustus* (2–1.5 million years ago) and in east Africa by *Australopithecus boisei* (2.2–1.2 million years ago).[1] A skull found west of Lake Turkana, the "black skull," represents a third robust species, *A. aethiopicus* (2.5–2.2 million years old). It is probably ancestral to *boisei*.[2] The robust lineage became extinct about 1 million years ago.

The gracile branch of *Australopithecus* was considered for a long time to contain two small-brained species: a northern species (*A. afarensis*) from Tanzania to Ethiopia (3.5–2.8 million years ago) and a southern species (*A. africanus*) in south Africa (3.0–2.4 million years ago). These two hominid species were walking bipedally but their relatively long arms and other characteristics indicate that they were still semiarboreal. Their brain was hardly larger than that of modern chimpanzees, and they were probably closer to the apes than to humans.

While anthropological researches continued, massive molecular evidence was confirming not only that the human species is closely related to the African apes but that, to everybody's surprise, the chimpanzee is more closely related to humans than it is to the gorilla; that is, the gorilla branched off from the chimpanzee line slightly earlier than the hominid line did.[3] Molecular evidence suggests that the branching of the human line from chimpanzees occurred as recently as 5 to 6 million years ago.[4]

In spite of numerous expeditions and diligent searching all over Africa, no species of australopithecine older than *Australopithecus afarensis* was found for many years. Then, in 1994, a species was discovered in Ethiopia that was determined to have lived 4.4 million years ago, close to the time when the hominid line branched off from the chimpanzee line. The study of this material has only just begun, but this fossil displays distinctly more similarities with chimpanzees than does *A. afarensis*. This Ethiopian fossil, named *Aridipithecus ramidus,* is now the oldest well-known fossil hominid. Following its discovery, foot bones and teeth older than *afarensis* and *africanus* were found in east and south Africa documenting the intermediate steps between *ramidus* and *afarensis/africanus*. No truly revealing remains have been discovered for the period between 8 and 4.4 million years ago.

It is very likely that the common ancestor of humans and chimpanzees was, like the chimpanzees, a knuckle-walker, and that each feature—ranging from the extremities, skull, brain, and teeth to the macromolecules—evolved at its own rate (a process called "mosaic evolution"). In other words, the "type" *Homo* did not evolve as a whole. Even today, humans and chimpanzees are extraordinarily similar in the structure of hemoglobin and other macromolecules, while in brain development and associated behaviors they differ profoundly.

THE RISE OF HOMO HABILIS, H. ERECTUS, AND H. SAPIENS

Between 1.9 and 1.7 million years ago the gracile australopithecines gave rise to a new species called *Homo habilis*. It is characterized by distinct skull features and increased brain size, and simple stone tools are present wherever *habilis* fossils are found. The *habilis* hominids were at first very puzzling owing to the seemingly great variation in body and brain size of the samples. It was eventually concluded that two species were involved, and the larger was renamed *Homo rudolf-ensis.*

Homo habilis is thought to be the ancestor of *Homo erectus,* a larger species with a considerably larger brain. Yet *Homo erectus* appears in the fossil record in Africa as far back as *habilis,* that is, about 1.9 million years ago, and the lifestyles of *habilis* and *erectus* may have been similar except that some populations of *erectus* seem to have tamed fire. *Homo erectus* was perhaps the first hominid that switched from a largely vegetarian to a partly meat-eating diet. In other words, this species became a scavenger and hunter.

Homo erectus was apparently highly successful and spread rapidly from Africa across the Near East to Asia, where his earliest remains were found in Java in strata about 1.9 million years old. *Homo erectus* shows some geographic variation, though it is puzzling how little evolution took place between the earliest *Homo erectus* and the latest ones (less than 300,000 years ago). Only simple stone tools are found with the earliest remains; more complex hand axes (bifaces) begin to show up in strata about 1.5 million years old but show virtually no advance in the ensuing million years. *Homo habilis* already had primi-tive stone tools 1.9 million years ago.

The species *Homo sapiens,* to which modern humans belong, some-how evolved from *Homo erectus,* but where and how has been highly controversial. There are two major theories of the origin of modern humans. One is that they evolved everywhere from local populations of *Homo erectus.* This multiregional theory was originally based on a putative similarity of the geographical races of modern *Homo sapiens* to the corresponding *Homo erectus* fossils in Africa, China, and the East Indies. This led Coon (1962) to the theory that a selection pressure for an enlarged brain throughout the wide range of *Homo erectus* in due time led to the gradual conversion of polytypic *Homo erectus* into polytypic *Homo sapiens.*

The opposing theory is sometimes called the "Mother Eve" hypothe-sis, and is based on mitochondrial reconstructions. According to this scenario, between 200,000 and 150,000 years ago a colonization wave of a new species originating in sub-Saharan Africa gave rise to all now-living populations of humans. This species (*Homo sapiens sapi-*

ens), it is claimed, arose from archaic *Homo sapiens* (themselves descendants of African *Homo erectus*) somewhere in sub-Saharan Africa less than about 200,000 years ago. *Homo sapiens sapiens* can be found in the Near East about 100,000 years ago; the East Indies, New Guinea, and Australia about 60,000 years ago; western Europe about 40,000 years ago (where their remains are known as Cro-Magnon Man); and the Far East at least 30,000 years ago. The skeleton of Cro-Magnon is quite similar to that of living humans and is considered to be in the same species. Cro-Magnon was responsible for the finest examples of cave art, at Chauvet, Lascaux, and Altamira, and for the finest stone tools.

In 1994 Ayala produced molecular evidence that would seem to refute the Mother Eve hypothesis and support the theory of regional continuity in human evolution from the time of *Homo erectus* to the present. He believes that the high frequency of ancient polymorphisms in the human gene pool precludes the possibility that the human species has passed through a narrow bottleneck, as claimed in the Mother Eve hypothesis.

Acceptance of a multiregional theory of human evolution would help to explain another puzzle in the fossil record: the fact that from China and Java to western Europe and even Africa, fossils of archaic *sapiens* have been found which are still very similar to *erectus* but with a larger brain (about 1200cc). These seemingly intermediate fossils are dated from 500,000 to approximately 130,000 years ago.

THE NEANDERTHALS AND CRO-MAGNON MAN

Ever since the finding of Neanderthal fossils in 1849 at Gibraltar, the relationship between *Homo sapiens* and Neanderthals has been the subject of continuing dispute. We know that between about 130,000 and 150,000 years ago in the West, long before the arrival of Cro-Magnon Man *(Homo sapiens sapiens),* archaic *sapiens* populations were replaced by Neanderthals, who ranged from Spain (Gibraltar) across Europe to western Asia (Turkestan) and south into Iran and Palestine (but not into Africa or Java). Neanderthals had a brain that on average was somewhat larger (up to 1600cc) than that of modern humans,

but they had only a primitive stone culture and showed no evolutionary change during the 100,000 years or so that they existed. The
Neanderthal branch of the hominid line became extinct around 30,000
years ago, or slightly less, long after Cro-Magnon types had invaded
Europe.

Were Neanderthal and Cro-Magnon two geographic races or two
different species? Owing to their great physical differences, they were
at first proclaimed to be two species. But then, based on the belief
that they excluded each other geographically, they were reduced to the
rank of geographic races (subspecies). They were again raised to the
rank of species when it was thought that Neanderthal and modern
humans had coexisted in Palestine in different caves in the same area
for a period of about 40,000 years (between 100,000 and 60,000 years
ago). But this was a period of great climatic fluctuations, and it was
eventually determined that Neanderthal lived in Palestine during the
coldest periods, while *H. sapiens sapiens* inhabited the area during
warmer and more arid spells. Hence, although both hominids are
found in the same region, on the whole they did not coexist at the
same place at the same time.

When Neanderthal was considered to be conspecific with modern
sapiens, some fossils from the Palestine caves were interpreted as
documenting interbreeding between the two types. This is not supported by a more recent analysis; and in spite of 10,000–15,000 years
of coexistence, no evidence for interbreeding has been found anywhere
in Europe. Neanderthal disappeared some 15,000 years after *Homo
sapiens sapiens* had invaded the area in Europe then occupied by
Neanderthal. In eastern and southern Asia archaic *sapiens* also eventually disappeared, to be replaced by modern *sapiens.*

CLASSIFYING FOSSIL HOMINID TAXA

Before the 1950s, the study of human origin was virtually the monopoly of anatomists, and hominid classification was dominated by
typological and finalistic thinking. There was little appreciation of the
uniqueness of individuals or of the enormous variation within a
species. Each fossil find was considered a different type and usually

given a binomial name, and these were considered members of a single ascending series connecting the primate ancestors with modern man.[5]

But the actual fossil finds did not confirm this conceptual construct nearly as well as one might have expected. Particularly disturbing was the sudden appearance of new types of hominids without any seeming connection with any preceding type. For instance, there is a huge gap between *Homo habilis* and its supposed ancestor, *Australopithecus africanus*, and between *Homo erectus* and its purported ancestor *Homo habilis*, and between *Homo sapiens* and its ancestor *Homo erectus*. Another incongruity was presented by the difficulty of placing geographically distant finds in the same linear sequence or vertical column.

Those who had adopted typological, one-dimensional thinking were obviously unaware of the widespread geographic speciation among the tetrapods. Most species of primates show geographic speciation, and most primate genera (except very large ones, such as lemurs and *Cercopithecus*) still consist of allopatric species. There is every reason to believe that the fossil hominid genera also consisted of allopatric species. This is confirmed by the restriction of *Australopithecus africanus* to south Africa and *afarensis* farther north, and of *Australopithecus robustus* to south Africa and *boisei* to east Africa.

The area from east to south Africa where most fossil hominids have been found is small, and founder populations of additional hominid species might have occurred in the vast areas of western, central, and northern Africa that have not yet been explored. (Indeed, australopithecine fossils 3.5–3.0 million years old have very recently been discovered in Chad, in central Africa.) Literally dozens of allospecies of *ramidus, afarensis, robustus, habilis,* and *erectus* could have lived in unexplored parts of Africa. The suddenness of some of the transition in the fossil record could be explained by "budding."[6] This means that the new descendant type originated somewhere in a peripherally isolated population and established contact with the parental species only after it had completed its genetic restructuring. The chance that we will ever discover the location of such an isolate is rather slight.

When only a few hominid fossils were known, it was easy to classify them into species: *afarensis, africanus, habilis, erectus,* and *sapiens.* Each of these names represented between 0.25 and 1.5 million years. In recent years numerous additional finds are either intermediate in time between typical specimens or from different geographical regions, hence not quite typical either. They usually show a good deal of mosaic evolution: some of their characters are those of their ancestors, some are those of the later descendants, while the rest are intermediate.

How to place living *Homo sapiens* in the taxonomic system has provoked the most astonishing disagreement, primarily the result of differences in the characters used from one classification system to the next. Julian Huxley (1942), basing his judgment on the uniqueness of the human species particularly with respect to its culture and domination of the world, suggested erecting a separate kingdom, Psychozoa, for *Homo sapiens.* Half a century later, Diamond (1991) went to the other extreme, by placing the chimpanzees in the genus *Homo* on the basis of their molecular similarity. While Huxley overemphasized human uniqueness, Diamond fell into the opposite error by ignoring it altogether.

One of the oldest axioms of classification is that characters should not merely be counted but also weighed. The accelerated evolution of the human central nervous system, the greatly expanded period of parental care, and all the physiological, social, and cultural developments that this permitted certainly justify placing the human species at least in a separate genus from *Pan* (chimpanzees), molecular similarities notwithstanding. By Diamond's criteria, *Australopithecus* would also become a synonym of *Homo,* and our nomenclature would no longer be able to reflect the degrees of difference among the different types of hominids.

There is now reasonable consensus as to the major types of fossil hominids and their relationship, but the detailed reconstruction of the polytypic superspecies of fossil hominids will be possible only after the discovery of more fossils, together with a consistent practice of population thinking. Such thinking began to enter physical anthropology around 1950, but even today *Australopithecus africanus* and *Homo erectus* are still widely considered types. How widespread these populations were and (in the case of *Homo erectus*) how much geographic variation they exhibited is often ignored by physical anthropologists, as well as the possibility of a number of peripheral isolates.

Becoming Human

What made the rise of humans possible, and in what sequence were human characteristics acquired? For a long time students of human evolution were comfortable with the following scenario. As Africa's climate became drier in the Miocene, many troops of human ancestors became isolated in more open landscapes, where walking was advantageous. The freeing of arms and hands encouraged tool use, and this, in turn, exerted a selection pressure on brain enlargement for the invention and skillful use of new tools. Thus, in this scenario, bipedalism is the key to humanization, via tool use.

Much recent evidence regarding the transition from apes to humans refutes this simple story. It is true that among mammals, consistent upright bipedal walking is unique to humans. There are jumping bipedal mammals, such as kangaroos and certain rodents, and some that can rise on their hind legs, like certain primates and bears, or occasionally even walk bipedally, like spider monkeys, gorillas, and particularly chimpanzees, but this is never their primary mode of locomotion.

Bipedalism alone cannot account for tool use, however, and tool use alone cannot account for the explosive growth of the human brain. The extensive use of tools by chimpanzees, although not always necessarily homologous with human tool use, suggests that tool use in hominids was well established before bipedalism evolved. And for almost 2 million years after the time the earliest human tools appear in the fossil record, there was little advance in tool technology. Also, bipedalism did not at first correspond to any noticeable increase in brain size. For the 2 or more million years that several species of australopithecines lived, they were bipedal, but by almost any other criterion these species were still apes. The ability to walk upright did not cause them to become humanlike in brain size, which was still quite small.

The early australopithecines were still semiarboreal, with their feet adapted for climbing and with their arms relatively longer than the arms of later fossil hominids and modern humans. Consequently, their babies had to be well developed at birth in order to be able to cling to the mother during her arboreal activities, just as the young of the various species of apes do today. Between 2.0 and 2.5 million years ago, however, a shift to complete terrestrial living freed the arms and hands of mothers for carrying babies, and this permitted a prolongation of the helpless stage of the newborn. This slower development in turn allowed for continued growth of the brain in early infancy, which is so characteristic of the human species. Thus the major impact of bipedalism was on mothering behavior, not on tool use.[7]

The indications are that these early diagnostic features of *Homo*—complete bipedalism and infant-carrying—evolved somewhere in a peripherally isolated population of australopithecines, and not in the australopithecines as a whole. No doubt it was facilitated by the availability of an appropriate ecological niche, but on this we presumably will never attain certainty.

Acquisition of upright walking by the ancestors of humans required a good deal of reconstruction of the locomotory apparatus. The shift from the arboreal and partial terrestrial bipedalism of the gracile australopithecines to the fully terrestrial bipedalism of *Homo erectus* was a period of greatly accelerated evolution, but upright carriage has still not yet been perfected, as indicated by the frequency in modern humans of back and sinus problems.

The australopithecines were largely vegetarians, as are chimpanzees today. The complete shift to terrestrial bipedalism when *Homo erectus* evolved led, it was once believed, to a shift from a vegetarian to an animal diet, that is, to hunting. Moreover, the strong teeth and heavy facial musculature of *Homo erectus*, relative to modern humans, led earlier authors to the erroneous view that he was a savage brute. Recent evidence based on a study of tooth wear and reinterpretation of so-called campsites does not confirm this scenario. Although an occasional animal seems to have been part of the diet of *Homo erectus*, as it is of modern chimpanzees, the hunting of large animals was apparently a late development in our history.

An intermediate stage was presumably a mixture of hunting and scavenging—the consumption of carcasses provided by large predators (lions, leopards, hyenas). Bipedalism was undoubtedly advantageous in permitting groups to follow herds of ungulates, which provided carcasses. There was also a great advantage in being able to carry babies along; a troop of hominids was no longer confined to a small territory containing the location of the helpless young, as many other mammals were. However, hominids do not have the resistance to ptomaine poisoning found among genuine scavengers, and it is therefore unlikely that any hominid was ever a primary scavenger. Trevino (1991) has presented persuasive evidence that early *sapiens* derived much food from grass seeds and wild cereals.

Nevertheless, the eventual shift to large-scale hunting probably played an important role in the humanization process. It encouraged the establishment of more elaborate base camps, and it required the planning of hunting forays, the development of hunting strategies, and the creation of more efficient weapons. Most importantly, many aspects of this new mode of living required an improved system of communication, that is, speech.

COEVOLUTION OF LANGUAGE, BRAIN, AND MIND

The australopithecines had a small ape-sized brain (400–500 cc), while *Homo erectus* had a distinctly larger one (750–1250 cc). But really large brains evolved only in the last 150,000 years—a fraction of the period since the hominid lineage separated from that of the chimpanzees. What selection pressure favored such an astounding explosion in the evolution of the human brain?

In addition to infant-carrying and hunting, major factors that favored an increase in brain size were the development of speech and the acquisition and generational transmission of culture that speech allowed. It would be futile to single out any one of these factors as the dominant one because they are all interconnected and contributed jointly.

Language does not exist among animals. To be sure, many species have elaborate vocal communication systems, but these consist of the exchange of signals; there is no syntax, no grammar. If one has only signals available, one cannot report the story of past happenings or make detailed plans for the future. For over 40 years various researchers have tried to teach chimpanzees language, but in vain. The animals have shown remarkable intelligence in acquiring a large vocabulary and have employed it to give the correct signals, but their system of communication is unable to convey any of the things only a language can convey.

There is a wide gap between the signaling of a chimpanzee (or any other kind of animal) and a genuine language. Linguists at one time thought they might find intermediate systems of communication if they studied the languages of the most primitive surviving human tribes. Alas, without exception, they all have highly complex, mature languages. Several scenarios have described how language could have evolved from a system of signals, but since we have no "fossil languages" to fill the gap, we will never have certainty.[8] The best way to shed light on the evolution of language is perhaps the study of the acquisition of language by children. Darwin was one of the pioneers in this endeavor. Such studies are now in progress by several psycholinguists, but they have to be done comparatively among children of languages with drastically different grammars.

The development of speech exerted a selection pressure not only on the nervous system but also on the entire vocal apparatus, of the larynx and adjacent areas of the respiratory system. Some evidence suggests that the vocal apparatus of the australopithecines would not have permitted proper speech. The *Homo* line, however, was preadapted for the development of speech owing to the low position of the larynx, the oval shape of the tooth row, the absence of large spaces between the teeth, the separation of the hyoid from the cartilage of the larynx, the general mobility of the tongue, and the vaulting of the palate. Neanderthals lacked some of these anatomical properties and are thus believed to have been inferior in the articulation of sounds.

Could the lack of the capacity to produce true language help explain why Neanderthals did not make better use of their brain, which was as large as that of modern humans of robust stature? Neanderthal culture was relatively primitive compared with that of later modern humans, as indicated by its simple stone tools. Neanderthals had no bows and arrows, no fishing equipment, and so on. But early *sapiens* had perhaps an equally impoverished culture. A good deal more research is needed to clear up this and other remaining uncertainties about the coevolution of language, brain, and culture.

When speech developed about 300,000–200,000 years ago in small groups of hunter-gatherers owing to a selective premium on improved communication, the situation was favorable for a further increase in brain size. However, around 100,000 years ago this increase came to a sudden halt, and from the time of the Neanderthals and early robust modern humans to the present, the human brain has stayed the same size. One would have expected continued brain growth in the 100,000 years preceding the development of agriculture, which occurred around 10,000 years ago. A Great Leap Forward in culture, as Diamond calls it, seems to have come about very rapidly during this period, yet it was not correlated with an equivalent jump in brain size or changes in other physical characteristics. Just why this should be so has been speculated about, but no convincing answer has been found.[9]

One factor in the halt of brain growth, perhaps, was the increase in the size of the troop. Primitive humans probably had a population structure similar to that of troops of chimpanzees or small tribes of hunter-gatherers. In these small groups, mortality would have been high, fewer troop members would have been reproductively successful, and gene flow would have been limited. All of these factors will favor a high rate of evolution under strong selection pressures, with the result that brain size might have increased rapidly.

Once larger troops became the norm among humans, the reproductive advantage of the presumably better-endowed leader would have been reduced, gene flow would have increased among all members of the troop, and those with smaller brains would have enjoyed better protection, longer survival, and greater reproductive success than they would have experienced had human group size remained small. In other words, the greater social integration of humans, while contributing enormously to cultural evolution, might have caused humans to enter a period of stasis in the evolution of the genome.

What light can evolutionary studies shed on the origin of the human mind? The study of mind has long been thwarted by semantic confusion, which has tended to restrict the term to the mental activities of humans. Researchers in animal behavior have now established that there is no categorical difference between the mental activities of certain animals (elephants, dogs, whales, primates, parrots) and those of humans. The same is true of consciousness, traces of which are found even among invertebrates and perhaps protozoans. Mind and consciousness do not form a demarcation between man and "the animals."

The human mind seems to have been the ultimate product of a concatenation of numerous miniemergences, in both our primate and hominid ancestors. There simply was no instantaneous emergence. A product of an unbelievably complex central nervous system, the mind arose very gradually although at highly unequal rates at different stages. The period when language evolved, permitting both improved communication and the evolution of culture, was surely a period of greatly accelerated emergence of mind.

One thing we have learned in the last 40 years is that evolution, though still remaining continuous, can advance by definite pulses, and that not all features of a system evolve at the same time or at the same rate. The transition from being still "nothing but an animal," as exemplified by the gracile australopithecines, to being that unique species, the modern human, was always gradual, but it has been characterized by drastic shifts in the rate of change.

Cultural Evolution

From the australopithecines to *Homo habilis*, through *Homo erectus* and archaic *Homo sapiens*, to modern *Homo sapiens sapiens* some 200,000 years ago, the physical characteristics of the hominid line underwent constant changes, leading to upright carriage, speech, and a large brain. It was long assumed that human *culture* experienced a parallel steady development. This, however, is not the case. For 85 percent of the existence of the hominids, there was no conspicuous advance in culture.

One of the most important developments in human cultural evolution was social integration in the hominid group. Among the primates, some species, such as orangutans, are more solitary, and others, such as chimpanzees and baboons, live in much larger social groups.

The shift by the time of *Homo erectus* to a strictly terrestrial mode of living was accompanied by an increase in group size. The evident advantages were better protection against predators, greater ability to cope with other competing groups of conspecifics, and an improved efficiency in the search for new resources, especially food.

As a result, the group-as-such became a target of selection, and many behavioral and physiological changes that facilitated the survival, prosperity, and reproductive success of the group as a whole would be favored by natural selection. These include the continuous sexual receptivity of females, the concealment of estrus, the development of menopause, expanded life expectancy, and other characteristics of modern humans not found in apes, not even in chimpanzees.

Undoubtedly, there was a great deal of fierce competition among neighboring groups and tribes, with the superior groups very often exterminating the inferior ones. The disappearance of Neanderthals from western Europe is still unexplained. They seem to have coexisted for 15,000 years with the Cro-Magnons, whose culture and communication were far more advanced, and one cannot exclude genocide as the explanation for the Neanderthals' extinction. This would not have been a new development within the anthropoid line. Several observations have been made in recent years of chimpanzee groups systematically exterminating neighboring, competing groups.

Among social animals, the benefits of cooperation are offset to some degree by the potential for conflict within the group itself, particularly when males compete for females. Some of the conflict inherent in larger troop size was mitigated in humans by a cultural trend toward monogamy and social stratification. "Superior" males presumably practiced polygyny within the troop, as is found in primitive human tribes and some modern cultures (such as Islam) even today. But for the most part monogamy became a means to ameliorate conflict, and marriage eventually became a strategy to cement connections between families that might otherwise be competitors.

Since marriage was a social contract, the dissolution of a marriage usually created considerable difficulties and was discouraged. Incest-avoiding rules were developed and enforced in most human societies, presumably to lessen conflict within the family and to vary the gene pool. A few cultures have practiced polyandry (in which women take multiple husbands), but much more frequently the family of the groom had to pay for their brides because they were an important addition to the work force of the groom's family. Striking differences in social structure, particularly with respect to sexual freedom and the role of women, can be found among the thousands of human societies that exist even today.

Throughout the hominid line, the family has been the foundation of group structure. Among modern hunter-gatherers we usually see a division of labor between men and women, with men being the hunters (providing proteins and fat to the diet) and the women being the gatherers (supplying carbohydrates and some protein in the form of nuts). The two sexes thus form a cooperative unit. However, there is cohesion not only within the core family (husband, wife, children) but also among members of the extended family (grandparents, siblings, cousins, uncles, aunts). The extended family is important not just for mutual help but also for cultural cohesion and transmission to the next generation. The breakdown of the extended family and even the core family is one of the basic roots of the cultural breakdown in inner-city slums.

As the size of the group increased, division of labor and job specialization became more important, further contributing to social stratification. Feudalism is the most extreme example. Specialization permitted humans collectively to occupy ever more various ecological niches. While most other species of organisms occupy only a single niche, the human occupies a large number.

Indeed, if one recognizes the existence of adaptive zones, as did Simpson and Huxley, then humans occupy an entire adaptive zone all by themselves. And if one feels that distinctness in the occupation of one's adaptive zone is correlated with taxonomic distinction, than Huxley was not entirely wrong in establishing a separate kingdom, Psychozoa, for humans, in spite of our genetic similarity to chimpanzees.

THE BIRTH OF CIVILIZATION

A momentous instance of punctuation in the evolution of human culture seems to have occurred in the transition from the hunter-gatherer stage to that of agriculture and animal breeding. This happened only about 10,000 years ago, and yet it had in many ways a more drastic effect on the human species and its role on earth than anything that had happened in the millions of preceding years of hominization. It was the beginning of civilization.

Permanent settlements were established around 10,000 years ago, some of sufficient size to be referred to as cities by archaeologists. These settlements encouraged a further division of labor and accelerated technological progress and particularly, in this century, medical progress. Cities permitted trade and an exploitation of nonrenewable natural resources, and most of all they brought about an intensification of agriculture, leading to rapid population growth.

Through these many cultural achievements humans have succeeded in becoming largely independent of the environment. We can now live from the Arctic to the Antarctic, and from the most humid tropics to the edge of the desert. Houses, clothing, transportation, and all sorts of machineries permitted this emancipation from local climates and from the kind of ecological specialization to which every other organism is subject. For all these reasons the human population explosion has not yet been hit by Malthus's prediction. However, this adaptive success was bought at the expense of much irreversible exploitation of natural resources and destruction of natural habitats.

Human Races and the Future of the Human Species

The division of modern humans into races and the biological status of these races have been controversial from Blumenbach on. In the days of slavery the comfortable view was widespread among whites that whites, blacks, and the mongoloid Asians were three different species. This view has been totally abandoned, but the number of human races recognized by different authors—ranging from five to well over fifty—suggests that arguments over the meaning of "race" have not been resolved.

Typological thinking is never enlightening in the study of life, but it has been most vicious and deleterious in the consideration of human races. Modern molecular research has revealed that all so-called human races are very closely related to one another and are simply variable populations. They often differ from one another in mean values for various physical, mental, and behavioral characteristics, but there is wide overlap of their curves of variation.

Undoubtedly there are racial characteristics. The longer two races have been separated, the greater will be their genetic differences. Populations within a race are more similar to one another than are races.[10] No one would mistake a sub-Saharan African for a western European or an eastern Asian, based on such superficial physical aspects as skin color, eye color, hair, shape of nose and lips, shape of skull, and stature. Genetics and molecular biology have added many other average differences or diagnostic characters. But when it comes to the psychological characteristics that really count, the role of genes is largely undetermined.

Most of the truly crucial characteristics usually attributed to human races have nothing to do with their genotypes but are ethnic, cultural properties. Races have been said to be friendly, cruel, intelligent, stupid, reliable, devious, industrious, lazy, suspicious, prejudiced, emotional, inscrutable, and what not. Indeed, almost any attribute a person may have has been claimed to characterize one or another human race. I am unaware of a scientific confirmation of any of these claims, though it is true that certain human populations have had rather well-defined cultural characteristics, for example, the New England Puritans, the Gypsies of Europe, and the black ghetto populations of American cities. In this field reliable facts are difficult to establish, since the scientific study of biological differences among races is frowned upon as apt to lead to racism.

The question is sometimes asked what chance there is for the human species to break up into several species. The answer is: None at all. Humans occupy all the conceivable niches that a humanlike animal might occupy, from the Arctic to the tropics. Whenever geographically isolated races have developed in the last 100,000 years, they interbred readily with other races as soon as contact was reestablished. There is far too much contact today among all human populations for any kind of effective long-term isolation that might lead to speciation.

Then, it is sometimes asked, could the now-existing species of humans evolve as a whole into a better new species? Could Man become Superman? Here again, one cannot be hopeful. To be sure, there is abundant genetic variation within the human genotype, but modern conditions are very different from the time when some populations of *Homo erectus* evolved into *Homo sapiens*. At that time, our population structure was that of the small troop, in each of which there was strong natural selection with a premium on those characteristics that eventually resulted in *Homo sapiens*. Furthermore, as in most social animals there was undoubtedly strong group selection.

Modern humans, by contrast, constitute a mass society, and there is no indication of any natural selection for superior genotypes that would permit the rise of the human species above its present capacities. Indeed, many authors claim that we are currently experiencing a deterioration of the human gene pool. Considering the high variability of the human gene pool, genetic deterioration of the species is not an immediate danger. What is far more frightening and threatening to the future of humankind is the deterioration of the value systems of most human societies (see Chapter 12).

But what about artificial selection for superior genotypes? Darwin's cousin Galton was the first to suggest that by proper selection one could and should improve mankind even further. Galton coined the term "eugenics." People from the far Left to the far Right at first readily endorsed this ideal, conceiving of eugenics as a way to lift the human species toward greater perfection. It is sadly ironic that this noble original objective eventually led to some of the most heinous crimes that humankind has ever seen. When it was interpreted typologically, it became racism, and eventually led to Hitler's horrors.

Only eugenic measures could bring about a drastic genetic "improvement" in the human race, but for many reasons this is impossible. First of all, we have no knowledge of the genetic basis of the nonphysical characteristics of present and future humans that we might choose to manipulate. Second, in order to be successful and well balanced, human society at all times has to consist of a mix of many different genotypes, but no one has any idea what the "right" kind of mix might be, or how to select for it. Finally, and most importantly, the steps that would have to be taken in order to implement eugenics are simply intolerable in a democracy.[11]

THE MEANING OF HUMAN EQUALITY

That no two individuals are alike is as true of the human population as of all other sexually reproducing organisms. Each individual is a different combination of morphological, physiological, and psychological characteristics and of the genetic factors that contribute to the shaping of these characteristics. There is no doubt as to the great plasticity of the human phenotype, particularly as far as behavioral characteristics are concerned, but genes also make a contribution to human behavior and personality. Some people are congenitally clumsy; others have a fantastic manual dexterity. Some people have a definite mathematical talent; others lack it to a lesser or greater extent. Musical ability has always been recognized as having a large innate component.

Indeed, few human characteristics can be found in which there is not a great deal of variation (polymorphism) in every human population. It is precisely this diversity that forms the basis for a healthy society. It permits a division of labor, but it also requires a social system that makes it possible for each person to find the particular niche in society for which he or she is best suited.[12]

Most people favor equality and agree that equality means equal status before the law and equal opportunity. But equality does not mean total identity. Equality is a social and ethical concept, not a biological one. Neglect of human biological diversity in the name of equality can only do harm; it has been an impediment in education, in medicine, and in many other human endeavors.

Great sensitivity and a high sense of justice are required to apply the principle of equality in the face of human biological diversity. As Haldane (1949) said so rightly, "It is generally admitted that liberty demands equality of opportunity. It is not equally realized that it also demands a variety of opportunities and a tolerance of those who fail to conform to standards which may be culturally desirable but are not essential for the functioning of society."

CHAPTER TWELVE

Can Evolution Account
for Ethics?

Perhaps no other area of human concern was shaken as drastically by the Darwinian revolution of 1859 as the theory of human morality. Before Darwin, the traditional answer to the question "What is the source of human morality?" was that it was God-given. To be sure, leading philosophers from Aristotle to Spinoza and Kant had thought about the correlated questions, "What is the nature of morality?" and "What morality is best suited for mankind?" Darwin did not challenge their conclusions about these deeper questions. What he did was to render invalid the claim of morality's God-given origin.

For this he used two arguments. First, his theory of common descent deprived Man of the special place in nature that had been attributed to humans not only by the monotheistic religions but also by philosophers. Nevertheless, Darwin agreed that, with respect to morality, there is a fundamental difference between humans and animals. "I fully subscribe to the judgment of those writers who maintain that of all the differences between Man and the lower animals the moral sense or conscience is by far the most important" (1871:70). Yet, since humans had animal ancestors, this difference now had to be explained in terms of evolution. To admit any discontinuous difference between humans and animals would have meant a saltation, and Darwin was unalterably opposed to such a process. He, the champion of gradualism, insisted that everything, even human morality, must have evolved gradually. Evidently, Darwin appreciated that a long time has elapsed (now estimated to be at least 5 million years) since the branching point of the human from the ape lineage, and this time interval provided sufficient time for humans to pass gradually through all the intermediate stages of ethical development.

Second, his theory of natural selection eliminated all supernatural forces from the workings of nature and implicitly refuted the assumption of natural theology that everything in the universe, including human morality, is designed by God and governed by His laws. After Darwin, philosophers had the formidable task of replacing a supernatural explanation of human morality with a naturalistic one. Much of the literature on the relation between ethics and evolution of the last 130 years has been devoted to a search for a "naturalistic ethics," and several volumes on the subject appear annually, 125 years after Darwin first posed the problem in 1871.

Some of these authors have gone so far as to express the hope that a study of evolution would give us not just insight into the origins of human morality but a *fixed* set of ethical norms. Leading evolutionists have adopted the more modest proposition that natural selection, directed at the appropriate target, would eventually lead to a human ethics in which altruism and regard for the common good would play a prominent role. Ethicists insist, and they are quite right to do so, that science in general, and evolutionary biology in particular, are not constructed to provide a reliable set of specific ethical norms. But it is important to add that a genuinely biological ethics which takes human cultural evolution as well as the human genetic program truly into consideration would be far more consistent internally than ethical systems which ignore these factors. Such a biologically informed system is not derived from evolution but is consistent with it.

Traditionally, ethics has been an area of conflict between science and philosophy. Ethics involves values, and scientists, so say most philosophers, should stick to facts and leave the establishment and analysis of values to philosophy. But scientists point out that new scientific knowledge about the ultimate consequences of human actions leads inevitably to ethical considerations. The current problems of the population explosion, the increase of atmospheric carbon dioxide, and the destruction of tropical forests are just a few examples.

Scientists feel that it is their duty to call attention to such situations and to make proposals for how to correct them. This inevitably involves value judgments. Very often our understanding of the processes of evolution, as well as other scientific data, enables us to make the ethically most appropriate choice when several options are available for action.

The Origin of Human Ethics

If natural selection only rewards self-interest, hence the egocentricity of every individual, how could any ethics develop that is based on altruism and on a sense of responsibility for the welfare of the community as a whole? T. H. Huxley's essay *Evolution and Ethics* (1893) was the source of much confusion on this issue. Huxley, who believed in final causes, rejected natural selection and did not represent genuine Darwinian thought in any way. Natural selection, as he conceived it, operates only on the individual, and this led him to conclusions that disproved for him any constructive contribution of natural selection to the greater good. It is unfortunate, considering how confused Huxley was, that his essay is often referred to even today as if it were authoritative.

But Huxley was right in perceiving dimly that the self-interest of the individual was somehow in conflict with the well-being of society. The principal problem of any naturalistic human ethics is to resolve the puzzle of the existence of altruistic conduct by basically egoistic individuals. Particularly challenging for a Darwinian is the problem of how natural selection could have contributed to altruism. Does not selection always reward those individuals who are completely selfish?

The long and heated debate of the last 30 years has revealed that when authors use the term "altruistic" they often mean different things. Clearly, it always implies helpfulness to some other person. But does the act always have to be detrimental to the altruist? If an animal utters warning calls to inform members of its group of the approach of a predator, it certainly endangers itself by attracting the attention of the predator. An altruistic act is usually defined as an act "that benefits another organism at a cost to the actor where cost and benefit are defined in terms of reproductive success" (Trivers 1985).

But altruism, as used in everyday language, does not always have to include danger or any kind of disadvantage. The philosopher Auguste Comte coined the term to mean concern for the welfare of others. For instance, if, on a walk, I help an old lady who has fallen, I perform an altruistic act without any danger to myself. The "cost" to me at most would be that I lose a minute of time. We all know warm-hearted generous people who rejoice in doing good deeds of all sorts. Are not those of their good deeds that involve no cost also altruistic? Is the small effort invested in a good deed a "cost" in any significant sense?

I contend that it does not represent the normal usage of the term "altruism" to restrict it only to instances involving potential danger or damage to the altruist. When trying to determine how natural selection could have favored the rise of altruism, it is important to make a distinction between these various kinds of behaviors.

Darwin already saw part of the answer, but only in recent years has it been fully understood that a person is the target of selection in three different contexts: as an individual, as a member of a family (or more correctly, as a reproducer), and as a member of a social group. In the case of the individual as target, only selfish tendencies will be rewarded by selection, as Huxley perceived. But in the other two contexts, selection may favor a concern for other members of the group, that is, altruism. The ethical dilemmas so often observed in human behavior cannot be understood without taking this three-sided context into consideration.

INCLUSIVE FITNESS ALTRUISM

A particular form of altruism quite widespread in animals, primarily species with parental care or those that form social groups consisting principally of extended families, is called inclusive fitness altruism. It entails defense of the offspring by the mother and sometimes by the father, a tendency to protect other close relatives from or warn them against danger, a willingness to share food with them, and other kinds of behavior evidently beneficial to the recipient but at least potentially harmful to the actor.

As was pointed out by Haldane, Hamilton, and numerous sociobiologists, these behaviors will often be favored by natural selection because they enhance the fitness of the genotype shared by the altruist with the beneficiaries of the behavior, namely, offspring and close relatives. Such behavior enhances the inclusive fitness of the altruist, it is said. Whenever the gene pool of the next generation is affected in this way by the contributions of some animals toward the survival of their close relatives, the process is referred to as kin selection.

Parental care is the most conspicuous example of the kind of altruism that enhances inclusive fitness. As long as the behavior results in an overall benefit for the genotype of the altruist, it is, strictly speaking, selfish rather than altruistic behavior. The literature of sociobiology contains literally hundreds of cases of seemingly altruistic acts which in reality enhance inclusive fitness and are therefore ultimately selfish, from the "point of view" of the genotype.

Inclusive fitness altruism is one of the major bones of contention in the current evolutionary literature. Some authors seem to think that all of human ethics boils down more or less to raw inclusive fitness altruism. Others think that when genuine human ethics evolved, it altogether replaced inclusive fitness altruism. My own position is somewhat intermediate. I discern many remnants of inclusive fitness altruism in the behavior of humans, such as the instinctive love of a mother for her children and the different moral stance we tend to adopt when dealing with strangers as compared with members of our own group. Most of the moral norms laid down in the Old Testament are characteristic of this heritage. Yet, it seems to me that inclusive fitness altruism is only a minor portion of human ethics as these systems exist today, consisting primarily of the love of a parent for a child.

Darwin was fully aware of the existence of inclusive fitness. Speaking of the sacrificial death of men with superior faculties in a human tribe, he stated (1871:161), "If such men left children to inherit their mental superiority, the chance of the birth of still more ingenious members would be somewhat better, and in a very small tribe, decidedly better. Even if they left no children, the tribe would still include their blood relations," who, as Darwin explained, have a similar genetic endowment.

Selection pressures that lead to the spread of inclusive fitness altruism occur not only in primitive peoples but in all social animals in which extended families are the nucleus of social groups. That social animals have a remarkable ability to recognize and favor their relatives is emphasized by Darwin again and again: "The social instincts never extend to all the individuals of the same species" (1871:85). How well developed this sensing of relationship is in certain animals has been excellently documented experimentally by Pat Bateson (1983).

RECIPROCAL ALTRUISM

Solitary animals such as leopards have fewer opportunities than social animals to acquire inclusive fitness altruism. In solitary animals it is mostly restricted to the behavior of a mother toward her offspring. The one seeming exception to the conclusion that solitary individuals are not altruistic toward anyone except offspring is in the case of reciprocal altruism, a mutually beneficial interaction among unrelated individuals. The cleaning fishes that free large predatory fishes of external parasites are a typical example. An alliance between two individuals in a fight against a third individual is another.

Actually, the term altruism is used here in a broad sense because the putative altruist always benefits at once or expects to do so in the long run. Such reciprocal interactions, particularly among primates, always imply a kind of reasoning: "If I help this individual in his fight, he will help me when I get into a fight." In other words, such behavior is basically egotistical rather than altruistic.

Reciprocal altruism is simply reciprocal benefits or exchanges of favors. These benefits are sometimes subtle, however, as when a philanthropist receives the approval, respect, and admiration of his fellow citizens in exchange for charitable donations, or when a scientist receives a Nobel, Balzan, Japan, Crafoord, or Wolff prize for outstanding contributions to his field. Rewarding individual achievements that benefit the larger group in the long run is very important to the betterment of our society. We take reward for achievement completely for granted in sport—only outstanding athletes get Olympic medals. But it should be remembered that all the great achievements of humankind have been made by a fraction of one percent of the total human population. Without rewards or recognition for outstanding accomplishments, our society would soon disintegrate, as happened to those Marxist societies that were organized around the principle of equal compensation for all.

But not all altruistic behavior leads to a reward. Certainly we know cases of altruistic acts where the altruist did not expect, and in fact did not even want, a reward of any sort. It has been claimed that reciprocal altruism, if practiced regularly, may facilitate acts of pure altruism where no reciprocation to the individual or close relatives is expected. Thus reciprocal altruism in our prehuman ancestors may be one of the roots of human morality.

THE EMERGENCE OF GENUINE ALTRUISM

Beyond inclusive fitness altruism and reciprocal altruism, both of which evolved through selection pressure on the individual, a far more important source of human ethics consists of the ethical norms and behaviors that have evolved through selection pressure on human cultural groups. Rather severe group selection has been going on throughout hominid history, as Darwin was fully aware.[1] In contrast with individual selection, group selection may reward genuine altruism and any other virtues that strengthen the group, even at the expense of the individual. As history has repeatedly illustrated, those behaviors will be preserved and those behavioral norms will have the longest survival that contribute the most to the well-being of the cultural group as a whole. In other words, ethical behavior for humans is adaptive.[2]

Most animal associations (groups) cannot serve as the target of selection. The exceptions are the so-called social animals, in which one finds cooperation. Of course, not all aggregations of animals are social groups. For example, schools of fish or the vast herds of migrating African ungulates do not qualify.

The human species offers an illustration par excellence of social

animals. Early hominid groups—enlargements of the original family—
were simply a continuation of the troop structure found in social
primates. The young females or the young males probably left the
troop and joined another one, but otherwise the group's behavior
reflected inclusive fitness altruism. For an extended family or small
troop to evolve into a larger, more open society, the altruism that
before was reserved for close relatives had to be extended to nonrelatives. We see rudiments of such genuinely altruistic behavior among
other primate groups—for instance, baboons—where an interchange
between unrelated individuals occurs.[3]

In the course of human evolution, some hominid individuals must
have discovered that an enlarged troop had a better chance of being
victorious in an encounter with another troop than a small group
consisting simply of an extended family. Perhaps a troop in possession
of a desirable cave, water hole, or hunting ground attracted outsiders
wanting to benefit from these advantages. It was of selective advantage
for the troop to be strengthened by such added manpower, even
though enlarging the group required that concern for the well-being
of others be extended to distant relatives or nonrelatives, that is,
beyond the range of inclusive fitness. Eventually, cultural norms of
behavior toward nonrelatives were established to counteract the basic
selfish tendencies of individuals in the troop and to impose on them
the constraints of an altruism that directly benefits the group as a
whole. Ultimately most individuals, whose well-being is closely connected with that of the group, benefit as well, though of course some
do not (as in the case of deaths during war).

The ability to apply group norms appropriately required the evolving reasoning capacity of the human brain. The coevolution of a larger
brain and a larger social group made two new aspects of ethical
behavior possible: (1) natural selection, working through group selection, could reward certain unselfish traits that benefited the group,
even some that were detrimental to a given individual, and (2) humans, with their new reasoning power, could deliberately choose ethical behavior over selfishness, rather than rely purely on instinctual
inclusive fitness. Ethical behavior is based on conscious thought that

leads to deliberate choices. The altruistic behavior of a mother bird is not based on choice; it is instinctive, not ethical. As Simpson (1969:143) characterized the situation, "Man is the only ethical organism in the full meaning of the word, and there are no relevant ethics except human ones." The adaptive shift from an instinctive altruism based on inclusive fitness to a group ethics based on decision-making was perhaps the most important step in humanization.

To be characterized as ethical, behavior requires the following conditions, according to Simpson (1969): (1) there are alternative modes of action; (2) the person is capable of judging the alternatives in ethical terms; (3) the person is free to choose what he judges to be ethically good. Thus ethical behavior clearly depends on an individual's capacity to foresee the results of his actions and to be willing to accept individual responsibility for the results. This is the basis for the origin and the function of the moral sense.

Ayala (1987) expressed more or less the same thought when saying that humans exhibit ethical behavior because their biological constitution determines the presence in them of the three necessary and jointly sufficient conditions for ethical behavior. These conditions are: (1) the ability to anticipate the consequences of one's own actions; (2) the ability to make value judgments; and (3) the ability to choose between alternative courses of action.

The difference between an animal, which acts instinctively, and a human being, who has the capacity for making choices, is the line of demarcation for ethics. Feelings of guilt, bad conscience, remorse, fear, or of sympathy and gratification that generally accompany the performance of actions subject to moral evaluation document the conscious nature of human unethical or ethical behavior. The capacity for ethical behavior thus is closely correlated with the evolution of other human characteristics, such as a great extension of the period of infancy and youth and thus of parental care, a trend toward enlargement of the hominid troop beyond the extended family, and the development of tribal traditions and culture (see Chapter 10). It is on the whole impossible in these correlated developments to determine what is cause and what is effect.

How Does a Cultural Group Acquire Its Particular Ethical Norms?

This question has been debated by philosophers from Aristotle, Spinoza, and Kant right up to modern times. The two most widely adopted pre-Darwinian answers were that moral norms were either God-given or the product solely of human reason (which was itself God-given).

Darwin himself worried about whether only those actions should be called moral or ethical that are the result of careful deliberation—that is, reason—or whether courageous or charitable acts that were done impulsively or "instinctively" should also be considered moral acts. He tended to consider the involvement of deliberation an important aspect of morality, as when he defined a moral being as "one who is capable of comparing his past and future actions or motives, and of approving or disapproving of them." Yet he also considered ethical acts to be a quasi-instinctive response to a "social instinct" found in all the social animals. This solution merely led to the follow-up question of how and why this social instinct evolved.

Bertrand Russell had a similar idea, but articulated it more concisely. He considered as "objectively right . . . [that] which best serves the interest of the group. A comparison of ethical norms throughout the world shows that those groups were most successful in which the self-interest of the individual was at least to some extent subordinated to the welfare of the community." Russell's statement comes nearer to a satisfactory answer than Darwin's, because it refers to the relative success of different human cultural groups. Some had moral norms that enhanced the probability of success—that is, the longevity—of the group; others had maladaptive moral norms that led to rapid extinction.

It is easy to imagine a scenario in which the particular value system of a cultural group might lead to its prosperity and numerical increase, and this might, in turn, lead to genocidal warfare against neighbors, with the victor taking over the territory of the defeated. In such a scenario, intragroup altruism and any other behaviors that strengthen the group relative to other groups would be rewarded by selection over the course of time, and any divisive tendencies within a group would weaken it and in due time lead to its extinction. Thus, the ethical system of each social group or tribe would be modified continuously by trial and error, success and failure, as well as by the occasional modifying influence of certain leaders.

What is moral and what is best for a group may depend on temporary circumstances. Wilson (1975) reminds us of the modifications in the value system of the Irish during the potato famine (1846–1848) and of the Japanese during the American Occupation after World War II. The great differences among tribes with respect to infanticide, sexual license, property rights, and aggressiveness document the plasticity of cultural ethical norms. Indeed, it might well be disadvantageous if all human societies had the same norms. A high birth rate is ethical in a primitive tribe with high infant mortality; by contrast, a restriction to one or two children is of great benefit not only for the group as a whole but even for the individual family in an overpopulated country. In a rural society it may be most beneficial if the extended family lives together, but this might lead to interminable strife under crowded urban conditions.

The rank of a given ethical norm also varies from culture to culture, depending on circumstances. The low ranking of human rights by the present Chinese government provides just one example. Our American negotiators, who seem to take it for granted that the entire world has only a single ranking of ethical norms, are unable to understand the Chinese attitude. Part of the moral indoctrination of the young is to teach them the ranking of norms within their particular culture.

Western philosophers have attempted to overcome this seeming ethical relativity by proposing various yardsticks by which to rank values. The Golden Rule is one such yardstick. The utilitarian suggestion that the norm should be judged by the degree to which it contributes to the greatest good for the greatest number is another. Truthfulness has long been accepted as an outstanding value, and justice is clearly a high-ranking ethical norm in the West, though no one seems to be able to agree on what is just and what is fair. In recent years it has been said that all those attitudes that give meaning to the individual life should be ranked high.

Much of what is considered moral depends on the size of the group with which one is associated. In primitive societies there is apparently an optimal size for a social group. When it gets too large the leaders seem to lose control over the group and the group splits up. This has been observed for tribes of South American Indians and for some social animals. When a group is too small, on the other hand, it becomes vulnerable to attacks by competitors. With the coming of agriculture 10,000 to 15,000 years ago, an increase in the size of the group beyond that of primitive tribes was favored: the availability of a good food supply permitted population growth, and a larger group could protect itself better against marauders. But as the group size expanded, new ethical conflicts arose. A change in values—for instance, greater emphasis on property rights—was inevitable.

As the size of human cultural groups grew, and particularly after urbanization and the origin of states, different social strata developed within a single society, each with a somewhat different set of ethical ideas. To what extent this is inevitable and perhaps even somewhat desirable is debatable. When it created gross inequities, as in most feudal societies, it sooner or later led to a revolution. The fight for democracy and for the principle of equality in the West was a reaction to the social inequities of the preceding era.

Within some societies, the values of individuals throughout the group are homogeneous, while in others subgroups differ in their moral norms. The disagreements about abortion, the rights of homosexuals, the rights of the terminally ill, and capital punishment illustrate the amount of dissension in modern American society, which is ethically quite diverse.

REASON OR RANDOM SURVIVAL?

So what can we conclude about how particular cultures acquire their moral norms? Are they the product of human reason or simply the result of random survival of those groups with the most adaptive ethical system? The enormous variety in the moral norms of primitive human tribes would indicate that many of the differences are simply due to chance. But when we compare the major religions and philosophies, including those of China and India, we discover that their ethical codes are remarkably similar, despite their largely independent histories. This suggests that the philosophers, prophets, or lawgivers responsible for these codes must have carefully studied their societies and, using their ability to reason on the basis of these observations, must have decided which norms were beneficial and which others were not. The norms proclaimed by Moses or by Jesus in the Sermon on the Mount were surely to a large extent the product of reason. Once adopted, such norms became part of the cultural tradition and were culturally inherited from generation to generation.

According to some authors, every ethical act of a human is exclusively a consequence of a rational cost-benefit analysis. According to others, it is a response to the quasi-instinctive disposition which Darwin called a "social instinct." In my opinion, the true answer is surely somewhere in between. Obviously, we do not rationally develop a special moral norm for every ethical quandary. In most cases we make our decision by automatically applying the traditional norms of our culture. Only when there is a conflict between several norms do we undertake a rational analysis.

But how does an individual within a culture acquire these traditional norms? What are the respective roles of "nature and nurture" in the development of the moral sense?

How Does the Individual Acquire Morality?

After the rise of genetics in this century, the question "Is the moral sense inborn or acquired?" achieved more and more prominence. The behaviorists and their followers believed that we are born tabula rasa, so to speak—that all of our behavior is the result of learning. The ethologists and particularly the sociobiologists, by contrast, tended to believe in a great deal of genetic programming. What is the evidence either group can provide to back up its claim?

The behaviorists can point to impressive evidence for the noninnateness of much of humankind's ethical disposition. Evidence for this claim is highly diverse, including: (1) the drastically different kinds of morality of different ethnic groups and tribes; (2) the total breakdown of morality under certain political regimes or after economic disasters;

(3) the ruthless and amoral behavior often displayed against minorities, particularly slaves; (4) the callous behavior exhibited in war, for instance the uninhibited shelling and bombing of civilian population centers; (5) the warping of the character of a child when deprived of a mother or mother substitute during a critical period in its infancy or when sexually abused.

This kind of evidence led the behaviorists and their followers to deny the existence of any inborn component and to believe that all moral behavior is the result of reasoning, based on conditioned responses to environmental stimuli. Their opponents favored an appreciable genetic component.

All the evidence accumulated during recent decades indicates that the values held by individual humans are the result of both inborn tendencies and learning. By far the largest part is acquired through observation of and indoctrination from other members of the cultural group. But individuals seem to vary greatly in their capacity to acquire the moral norms of their group. This inborn capacity for acquiring ethical norms and adopting ethical behavior is the crucial contribution of inheritance. The greater such a capacity is in an individual, the better equipped he or she will be to adopt a second set of ethical norms supplementing (and in part replacing) the biologically inherited norms based on selfishness and inclusive fitness.

Some individuals from childhood on seem to be nasty, ruthless, selfish, dishonest, and so on. Others seem to be little angels from the start—warm-hearted, utterly unselfish, always reliable and cooperative, honest to the core. Modern twin and adoption studies document a considerable genetic component in these different tendencies. The researches of child psychologists have also revealed variation in personality traits in newborn and very young babies. Most of these traits do not change during adolescence.[4]

To demonstrate the heritability of a given trait is usually rather difficult. Curiously, it seems more easily demonstrated for bad traits than for good ones. Darwin cited the occurrence of kleptomania through several generations in highly affluent families as evidence for strict heritability of certain unethical behaviors. A genetic predisposition can often be inferred also in the case of psychopaths. Moreover, the universality of aggressiveness in all territory-holding animals and in nearly all primates (perhaps least in the gorilla) leaves little doubt that humans have an inborn tendency toward aggressiveness. The appalling rate of murders, domestic abuse, and other acts of violence among humans is sad confirmation of this heritage. Yet, as Darwin rightly said, "If bad tendencies are transmitted, it is probable that good ones are likewise transmitted" (1871:102).

But heredity is not all. Analysis of the birth-order effect has shown how pliable certain other character traits are, including leadership, creativity, conservative tendencies, and so on.[5] It will require much further research before human "moral" traits can be properly partitioned into those that are strongly innate and those that are largely acquired after birth.

AN OPEN BEHAVIOR PROGRAM

Both an innate ethical predisposition to acquire the norms of a culture (the contribution of "nature") and exposure to a set of ethical norms (the contribution of "nurture") are required if a child is to grow up with respect for the ethical system of his culture. Numerous studies have concluded that ethical norms are largely acquired during infancy and youth. I am rather persuaded of the validity of Waddington's thesis (1960) that a very special kind of learning is involved, a type akin to the imprinting of animals, so well described by the ethologists and illustrated by the attachment of a young gosling to its mother.

Humans are distinguished from all other animals by the extent to which their behavioral program is open. By this I mean that many of the objects of behavior and the reactions to these objects are not instinctive, that is, not part of a closed program, but are acquired in the course of life. Just as the gestalt of the mother goose becomes imprinted in the behavior program of the gosling after hatching, so in human beings ethical norms and values are laid down in the open behavior program of the infant. The enlargement of the brain and its storage capacity permitted a limited number of fixed reactions to the environment to be replaced by a capacity to store a large number of learned behavior norms. This provides for much greater flexibility and allows for more precise fine tuning. As Waddington proposed it, "The human infant is born with probably a certain innate capacity to acquire ethical beliefs, but without any specific beliefs in particular" (1962:126).

Darwin was fully aware of the power of imprinting in early youth: "It is worthy of remark that a belief constantly inculcated during the early years of life, while the brain is impressible, appears to acquire almost the nature of an instinct." This power of indoctrination, says Darwin, leads not only to the adoption of ethical norms but also to the unquestioned acceptance of certain "absurd rules of conduct" found in many human cultures (1871:99–100).

Psychologists who study learning have demonstrated that certain things are learned much more readily than others. An olfactory animal learns smells much more easily than a visual animal, and vice versa. It is distinctly possible that if certain moral norms had contributed to the survival potential of certain groups during hominid history, this would also favor the selection of a structure in the open program facilitating the storage of such behavior norms. In what part of the brain this information is stored and how it is retrieved under the appropriate circumstances are as yet unknown.

Every child psychologist knows how eager children are to receive new information, including normative rules, and how ready they are on the whole to accept them.[6] The value system of a person is largely controlled by what was incorporated in his or her youth into this open behavioral program. It is the vast capacity of this open program in humans that makes ethics possible. And the foundation laid in infancy lasts, under normal circumstances, throughout life.

If Waddington's thesis is correct, it follows that early ethical education is of the utmost importance. We have just passed through a period in which exaggerated importance was placed on the so-called freedom of the child, allowing it to develop its own goodness. We have made fun of moralizing in children's books and have tended to remove most moral education from the schools. This may cause few problems when parents perform their roles properly, but it may spell disaster when parents fail to do their job. In view of our better understanding of the origin of the morality of the individual, would it not seem time to again place greater stress on moral education? It is particularly important that such education be started at the earliest possible period. It is the young child who is most willing to accept authority and is most easily impressed by norms. Half an hour of ethical education a day in elementary school might have a major impact. It would be far less effective to offer courses on ethics in college, as has recently been urged by a university president.

We are living in a time of changing values, and most members of the older generation deplore a breakdown of morals. If someone were to claim that this breakdown is to a large degree due to the deterioration of the ethical instruction of our youth, it would be difficult to refute such a claim. A good ethical education strengthens the awareness that one is responsible for one's own actions, by training a person to ask himself, from childhood on, whether his behavior conforms to the highest standards of the society. The powerful constraint on behavior exerted by this self-examination is usually referred to as one's conscience.

A great deal of ethical literature being written today is pessimistic, if not despairing. The genetic determinists are so impressed by the evil, aggressive heritage of humans that they despair that any period will ever arrive in which humanity's good heritage acquires again the upper hand over "Das sogenannte Böse [the so-called evil]," as Lorenz (1966) expressed it. In the other camp, psychologists and educators who believe in the dominance of environmental influences over inheritance are frustrated by the fact that good ethics are not acquired in spite of the most rational presentation of the subject. But they ignore Waddington's theory that ethical norms have to be acquired from earliest childhood on, by an imprinting-like process, and that such instruction has to be incessant. How successful moral education can be is documented by the low crime rate among many religious communities, such as Mormons, Mennonites, Seventh Day Adventists, and others. All one would have to do to improve the situation drastically is to step up ethical instruction and begin it at the earliest possible age.

Quite a few readers will smile at such seemingly old-fashioned advice. Is that the best science can come up with, they might say? Let me be clear that I am really quite serious. I have looked at school books, I have looked at children's stories, and I have looked at quite a few television programs. More often they are designed to entertain and—so far as they are educational—to transmit information in the most painless manner. Does one encounter moral instruction? Occasionally, on public broadcasting programs. But all too rarely. Why? Because, one may be told, brainwashing a child is an interference with its personal freedom, or moralizing is not entertaining and therefore will not sell. Personally, I do not see how a high standard of ethical behavior can be achieved in a culture if the will to implement it is absent.

What Moral System Is Best Suited for Humankind?

The major problems that humanity has traditionally confronted, such as wars, disease, and food shortages, are being addressed with increasing success as we approach the millennium. By contrast, a series of other problems are rising in significance, problems which ultimately have to do with values. Among these are the breakdown of the family, the drug problem, domestic abuse and other acts of violence, a decline of true literacy (along with a growing addiction to television, video games, and professional sports), uninhibited reproduction, waste and depletion of natural resources, and destruction of the environment. Are the traditional ethical norms of the Western world adequate to help us deal with these present and future problems?

The traditional ethical norms of Western culture are those of the Judeo-Christian tradition, that is, they are based on the various commandments and injunctions articulated in the Old and New Testaments. As phrased in the sacred texts, such commandments would seem to be absolute, allowing for no deviation. Ordinarily the commandment "Thou shalt not kill," for example, has absolute validity. However, to withdraw life-support machinery from an intensely suffering terminal patient is an act of mercy. And a similar flexibility must be applied in the case of abortion. When an unwanted child would have to face a life of misery and neglect or when his mother would be led to total despair, then abortion would certainly seem to be the more ethical option. And it is nonsense to drag the question of life into this argument, for as a biologist I know that every egg and spermatozoon also has life.

There are two reasons why the traditional norms of the West are no longer adequate. The first is their rigidity. At the very core of human ethics is the possibility of making a choice and of evaluating the conflicting factors in order to make the right choice. As much as ethical norms are part of our culture, the responsibility for applying them rests with the individual; if the norms are too rigid, the individual may have to make the decision not to adhere. It is also important to remember that the essence of the evolutionary process is variability and change; hence, ethical norms must be sufficiently versatile to be able to cope with a change of conditions. Ethical decisions often depend on the context. Absolute prescriptions rarely solve ethical dilemmas and can in certain circumstances be highly unethical if followed inflexibly. Moreover, depending on the circumstances, there is often a pluralism of possible solutions, and the best outcome may invoke a combination of different solutions.

The second reason is that humankind has indeed experienced a drastic and accelerating change of conditions. The ethical norms adopted by the pastoral people of the Near East more than 3,000 years ago are being proven inadequate for the modern urbanized mass society in a greatly overpopulated world. The set of moral standards that would most benefit strictly territorial, pastoral people is very different from the set that would be adaptive in the huge urban centers of today. As Simpson (1969:136) has rightly said: "All ethical systems that originated under tribal, pastoral, or other primitive conditions have become, to greater or lesser degree, [mal]adaptive in the extremely different social and other environmental conditions of the present time."

It seems to me that at least three broad ethical problems facing our modern world might not be adequately covered by the West's traditional ethical norms. The first is what Singer (1981:111–117) refers to as the problem of the "expanding circle." Not only in primitive societies but also in the Old Testament, among the Greeks, and even among eighteenth- and nineteenth-century Europeans in Africa and Australia, a totally different ethic was employed toward outsiders than toward members of one's own group. In the United States whites exhibited this behavior toward blacks, particularly in the South, until a few decades ago, and apartheid in South Africa was a contemporary remnant of such group selfishness. Even within ethnically homogeneous societies, such as early twentieth-century Britain, there are or have been differences about minor virtues, loyalties, and injunctions among religious groups, political parties, professional groups, social stations, and so on. Such differences set up tensions and conflicts. This is particularly true if the more articulate upper classes are responsible for codified norms of ethics, which to some extent may be in conflict with the morality of lower socioeconomic strata. The rebellion of the early Christians against the morality of the decaying Roman Empire well illustrates this situation.

As the circle of one's group expands and groups with different ethical systems fuse, inevitably there are conflicts, because each group is convinced of the superiority of its own moral values. To appreciate this problem, one only has to think of the difference in moral stance of a modern American and an Islam fundamentalist with respect to the rights of women, or, in our own country, the different attitudes toward abortion shown by certain religious groups versus feminist groups. In spite of the difficulties, the ethics of the future must address the problem of how to proceed when one's own values clash with those of another group.

The second great ethical problem of our time is excessive egocentricity and attention to the rights of the individual. "Expanding the circle" in our own society has resulted in a legitimate fight for equality, particularly by minorities and women, but this has also had some undesirable side effects. Martin Luther King, Jr., has been perhaps the only freedom fighter who has reminded his followers that all rights must be accompanied by obligations. Our excessive narcissism has many roots—mass society, the teachings of Freud, reaction to the preceding disregard for the rights of the individual, a political system that depends on the politician's appeal to the individual voter, and the emphasis of monotheistic religions on individual ethics. Almost invariably crucial dilemmas arise when a choice must be made between individual ethics and social or community ethics. This can be seen in such controversies as birth control, taxes for improving our environment, and humanitarian aid for hopelessly overpopulated countries.

The third great ethical problem of our day is posed by the discovery of our responsibility for nature as a whole. Growth, both of the economy and the population, used to rank very high in our Western value system. Even though certain influential people such as the late Nobel Prize–winning economist F. Hayek and the current Pope have so far failed to appreciate the danger of overpopulation, I cannot see how it can be ignored any longer. Certain of our societies, such as China and Singapore, have courageously tackled this problem by a reordering of ethical values, despite the concurrent loss of certain individual rights which many Western humanitarians have deplored. The sooner other overpopulated countries follow the example of China and Singapore, the better it will be for them, for our entire species, and for the planet on which we live.

The dilemma we are facing is the conflict between traditional and newly discovered values. The right to unlimited reproduction and exploitation of the natural world is inconsistent with the needs of human posterity as well as the right to existence of millions of threatened species of wild animals and plants. Where is the proper balance between human freedom and regard for the welfare of the natural world?

The concept that humankind has a responsibility toward nature as a whole is an ethical notion that seems to have originated remarkably late. It is curiously absent from most religions and other ethical codes. In recent times Aldo Leopold, Rachel Carson, Paul Ehrlich, and Garrett Hardin have been articulate in the United States in their championship of a conservation or environmental ethic. But much of what these modern Americans consider ethically valuable is in conflict with the immediate benefit of certain private individuals and is therefore resisted. And yet if the human species and the natural world as a whole are to have a future, we must reduce the selfish tendencies in our current value system in favor of a higher regard for the community and for the whole of creation. This requires a rejection of the ideal of continued growth, and its replacement with the ideal of a steady-state economy, even if this were to entail a reduction in our standard of living. The shift from a pastoral or agricultural society to an urban mass society requires considerable adjustments in our values, and so does the shift from a thinly settled world to the modern industrial world, with its massive overpopulation and giant cities. The ethical norms of the future must be flexible enough to evolve as these problems appear, if we are to remain an adaptive species.

The basic premise of the new environmental ethic is that one should never do anything to one's environment (in the widest sense of the word) which would make life more difficult for future generations. This includes the reckless exploitation of nonrenewable resources, destruction of natural habitats, and reproduction beyond the replacement value. This principle is very difficult to enforce because it is inevitably in conflict with selfish considerations. It will require a long period of education for all of humankind to understand this environmental ethic. Such education should start among young children, whose seemingly natural interest in animals, their behavior, and their habitat can be used to strengthen environmental values.

Is there any particular ethics that an evolutionist should adopt? Ethics is a very private matter, a personal choice. My own values are rather close to Julian Huxley's evolutionary humanism. "It is a belief in mankind, a feeling of solidarity with mankind, and a loyalty toward mankind. Man is the result of millions of years of evolution, and our most basic ethical principle should be to do everything toward enhancing the future of mankind. All other ethical norms can be derived from this baseline."

Evolutionary humanism is a demanding ethics, because it tells every individual that somehow he or she shares a responsibility for the future of our species, and that this responsibility for the larger group should be just as much a part of cultural ethics as concern for the individual. Every generation is the current caretaker not only of the human gene pool but indeed of all nature on our fragile globe.

Evolution does not provide us with a codified set of ethical norms such as the Ten Commandments. Yet evolution did give us a capacity for stretching beyond our individual needs to take those of the larger group into account. And an understanding of evolution can give us a worldview that serves as the basis for a sound ethical system—an ethical system that can maintain a healthy human society and provide for the future of a world preserved by the guardianship of man.[7]

Notes / Bibliography / Glossary

Guide to Topics Covered

Acknowledgments / Index

Notes

1. What Is the Meaning of "Life"?

1. The search becomes even more futile if the words "mind" or "consciousness" are substituted for "life." This substitution was made to facilitate a demarcation between human life and the life of animals, but it turned out to be a poor strategy because there is no definition of either mind or consciousness that would be applicable only to humans and exclude all animals.

2. Many attempts were made in the last century to define living or life in a simple sentence, some of them based on physiology, others based on genetics, but none of them is completely satisfactory. What has been successful is an ever more correct and ever more complete description of all aspects of living. One might say that "living consists of the activities of self-constructed systems that are controlled by a genetic program." Rensch (1968:54) states, "Living organisms are hierarchically ordered open systems composed prevailingly of organic molecules, consisting normally of sharply delimited individuals composed of cells and of limited timespan." Sattler (1986:228) says that a living system can be defined "as an open system that is self-replicating, self-regulating, exhibits individuality, and feeds on energy from the environment." These statements are more descriptions than definitions; they contain statements that are not necessary, and omit references to the genetic program, perhaps the most characteristic feature of living organisms.

3. The historians Maier (1938) and Dijksterhuis (1950, 1961) have splendidly described the gradual change from the Greeks through the "Dark Ages" to scholastic philosophy and, finally, to the beginnings of the Scientific Revolution, indicated by the names Copernicus, Galileo, and Descartes. These historians have determined the manifold influences on this development and what it had retained from the Greek tradition. This includes, for instance, "the passionate endeavor of classical physical science to trace the immutable everywhere behind the variability of phenomena" (Dijksterhuis 1961:8), that is, essentialism. "The fundamental idea

of [Plato's] entire philosophy is that the things perceived by us are only imperfect copies, imitations or reflections of ideal forms or ideas" (Dijksterhuis 1961:13). In the developments of these ideas, Plato evidently had more influence than Aristotle. It was he who "wholeheartedly endorsed the Pythagorean principle . . . as the germ of the mathematization of science." "Plato makes the cosmos a living being by investing the world-body with a world-soul" (Dijksterhuis 1961:15).

4. Actually, this is a rather simplistic presentation of the pathway by which Descartes arrived at his conclusion. The story goes back to the Aristotelian teaching, accepted by the scholastic philosophers, that plants have a nutritive soul and animals a sensitive soul, only man having a rational soul. Material substance was credited to the sensitive soul of animals, while the rational soul was immortal. The capacities of the sensitive soul of animals were limited to sensory perceptions and memory. From his discussions it is clear that Descartes understood under (rational) soul the "reflective consciousness of self and of the object of thought." To ascribe the capacity of rational thought to animals would credit them with an immortal soul, and that for Descartes was an unacceptable proposition, because it meant that their souls would go to heaven. (The atheistic thought that perhaps there was no heaven for human souls either apparently never occurred to Descartes.) Ultimately, Descartes's reasoning was based on scholastic definitions of substance and essence, which precluded the existence of soul in animals, and limited it to thinking, rational humans. This conclusion eliminated the unacceptable possibility of animals having an immortal soul rising to heaven after death (Rosenfield 1941:21–22). If one could deny a soul to animals, it was self-evident that one could not accept the belief still widespread in Europe in the seventeenth century that a soul, an anima or *vita mundi,* permeated the universe.

5. The word mechanist was used throughout the nineteenth and part of the twentieth centuries in two different meanings. On one hand it referred to the views of those who denied the existence of any supernatural forces. For the Darwinians, for instance, it meant the denial of the existence of any cosmic teleology. However, for others the major meaning of the word mechanist was a belief that there is no difference between organisms and inanimate matter, that there is no such thing as life-specific processes. This was the major meaning of mechanist for the physicalists.

6. Nägeli (1845:1) says that it must be postulated of the specific terms used in an explanation "that they are expressed generally, absolutely, and in the form of a movement." Rawitz (fide Roux 1915) defines life as "a special form of molecular movements and all manifestations of life are variants of it."

7. Most existing histories of vitalism are rather one-sided, having been written either by vitalists such as Driesch (1905) or by their opponents, who saw nothing good in it. Perhaps the best account is that of Hall (1969, chaps. 28–35). Blandino's (1969) treatment concentrates on Driesch; and Cassirer (1950) likewise focuses on Driesch, his followers and opponents. Jacob's (1973) concise presentation is well balanced and follows the fate of vitalism from animism up. An even more comprehensive and truly balanced history of vitalism, however, is still lacking.

8. As Lenoir (1982) has correctly pointed out.

9. "In fact, various forms of vitalism represent quite legitimate extensions of the Cartesian program in mechanistic biology with Newtonian means" (McLaughlin 1991).

10. How similar Müller's concept of the Lebenskraft was to the concept of the genetic program may be documented by a few quotations: "[Müller's] Lebenskraft acts in all organs as cause and supreme effector of all phenomena according to a definite plan [program]" (DuBois-Reymond 1860:205). Parts of the Lebenskraft, "representing the whole, are transmitted at reproduction without incurring any loss to every germ where it can remain dormant until germination" (ibid.). The four principal attributes of the Lebenskraft listed by Müller are indeed characteristics of the genetic program: (1) not being localized in a specific organ, (2) being divisible into a very large number of parts all of which still retain the properties of the whole, (3) disappearing at death without leaving any remnant (there is no departing soul), and (4) acting according to a plan (having teleonomic properties). I have described the beliefs of J. Müller in considerable detail, to correct the whiggish treatments by such physicalists as DuBois-Reymond, who maligned Müller as an unscientific metaphysician.

11. Von Uexküll, B. Dürken, Meyer-Abich, W. E. Agar, R. S. Lillie, J. S. Haldane, E. S. Russell, W. McDougall, DeNouy, and Sinnott, to mention only a few of the numerous early twentieth-century vitalists. Ghiselin (1974) designates W. Cannon, L. Henderson, W. M. Wheeler, and A. N. Whitehead as cryptovitalists.

12. Goudge (1961), Lenoir (1982). An opposition to Darwin's selectionism was also a frequent component of vitalist arguments (Driesch 1905).

13. Beginning with C. F. Wolff (1734–1794) the idea developed that there was a basic, undifferentiated stuff which gave rise to the more formed elements. F. Dujardin (1801–1860) first described (1835) and defined it properly under the name "sarcode." More and more attention was paid to it as microscopy flourished. Purkinje coined the term "protoplasm" in 1840. In 1869 protoplasm was for T. H. Huxley the physical basis of life. Cytoplasm was the term introduced by Kölliker to designate the cellular material outside the nucleus.

14. Actually, this term had been used in the social sciences all the way back to Comte, although organicism meant something rather different for the sociologists from what it meant for the biologists. Bertalanffy (1952:182) listed some 30 authors who had declared their sympathy for a holistic approach. This list was very incomplete, however, not even including the names of Lloyd Morgan, Smuts, and J. S. Haldane. F. Jacob's (1973) concept of the integron is a particularly well argued endorsement of organismic thinking.

15. Woodger (1929) gives an impressive list of biologists who endorsed the organicist viewpoint. E. B. Wilson (1925:256), for instance, said "even the most superficial acquaintance with the cell activities shows us that [explaining the cell as a chemical machine] cannot be taken in any crude mechanical sense—the difference between the cell and even the most intricate artificial machine still remains too vast by far to be bridged by present knowledge . . . modern investigation has brought ever-increasing recognition of the fact that the cell is an organic system, and one in which we must recognize the existence of some kind of ordered structure or

organization." Not surprisingly, holistic thinking was always particularly well represented among the developmental biologists. It was strong in the writings of C. O. Whitman, E. B. Wilson, and F. R. Lillie. Haraway (1976) devotes the major part of an entire book to the organicism of three embryologists, Ross Harrison, Joseph Needham, and Paul Weiss. Interestingly, Harrison considered emergence a metaphysical principle, and therefore he considered Lloyd Morgan a vitalist. Like so many biologists in the post-1925 period, he thought that the newly discovered principles of physics, such as the relativity theory, Bohr's complementarity principle, quantum mechanics, and Heisenberg's indeterminacy principle, applied equally to biology and physics.

16. Nagel (1961) defined a mechanist in biology as "one who believes that all living phenomena can be unequivocally explained in physico-chemical terms, that is, in terms of theories that have been originally developed for the domains of enquiry in which the distinction between the living and non-living plays no role, and that by common consent are classified as belonging to physics and chemistry." Such reduction characterizes all of Nagel's account.

17. For instance: "Holism is a specific tendency, with a definite character, and creative of all characters in the universe, and thus fruitful of results and explanations in regard to the whole course of cosmic development" (1926:100). Not surprisingly, holism as presented by Smuts was widely considered to be a metaphysical concept.

18. The subject of levels of integration was discussed in great detail in a special symposium volume (R. Redfield 1942).

19. One mistake that was made particularly often was to consider each level of integration to be a global phenomenon. This is not what these levels are. Every integron from the molecular to the supraorganismic level is singular. There is no conflict between this interpretation and Novikoff's (1945) statement that "the laws describing the unique properties of each level are qualitatively distinct, and their discovery requires methods of research and analysis appropriate to the particular level," and we would now add, appropriate to the particular integron. A modern evolutionist would say that the formation of a more complex system, representing a new higher level, is strictly a matter of genetic variation and selection. There is no conflict with the principles of Darwinism.

2. What Is Science?

1. This literature began with Whewell (1840) and led to the classical accounts of Nagel (1961), Popper (1952), and Hempel (1965), as well as the more recent volumes of Laudan (1977), Giere (1988), and McMullin (1988); much additional literature is listed in these volumes. All of these authors and many others have attempted to provide a definitive answer to this question. Pearson (1892) felt that what characterized science was sharing in the same methodology. But this criterion omits the important consideration that all true sciences, as we shall see, share also in certain principles, such as that of objectivity.

2. Nagel (1961:4). It is clearly easier to describe what science is and what scientists

do than to provide a concise and universally acceptable definition. Examples of descriptions are: "Science studies things that are puzzling and thus appeal to human curiosity"; or "The functions of science are prediction, control, understanding, and the discovery of causes" (Beckner 1959:39); or "Science is the organization and classification of knowledge on the basis of explanatory principles" (Nagel 1961:4). Other definitions are: "Science is the endeavor to increase our understanding of the world on the basis of explanatory principles and with a continuing critical testing of all findings" (Mayr ms.); or "Empirical science has two major objectives: to describe particular phenomena in the world of our experience and to establish general principles by means of which they can be explained and predicted" (Hempel). Others: "Science comprises all activities of human intelligence depending entirely on objective data and logic; also an unlimited testability of theories"; or science is "logically general sentences that are directly or indirectly open to observational confirmation and refutation, and can be employed in explanations and predictions"; or it is "the organization and classification of knowledge on the basis of explanatory principles."

3. For a detailed discussion of the nature of scientific problems see Laudan (1977).
4. See Hall (1954).
5. See Mayr (1996).
6. The German philosopher Windelband (1894) distinguished between two kinds of sciences, nomothetic and idiographic sciences, with the term science used in the German meaning of Wissenschaft (including the humanities). His terminology was meant to separate the natural (nomothetic) sciences from the (idiographic) humanities. Again this claim turned out to be invalid because biology was completely left out of his classification. His characterization of idiographic sciences, as dealing with unique and nonrecurring phenomena, was meant for the humanities, but this description fits many of the natural sciences, particularly evolutionary biology, equally well, as was correctly pointed out by Nagel (1961:548–549). By now we understand that the contrast between "science" and the humanities is not nearly as stark as Snow and Windelband thought. This new insight is the result of a number of considerations: (1) What both the physicalist philosophers of science as well as the humanists traditionally had designated as "science" was in reality physics, only one of the sciences, (2) the erosion of strict determinism and of the belief in the overwhelming importance of universal laws has made the contrast between science (even including the physical sciences) and the humanities much less absolute, (3) considering biology, particularly evolutionary biology, a part of science establishes a bridge between the natural sciences and the humanities, and (4) historical processes, so singularly neglected in most of the physical sciences, are eligible for scientific analysis and must be included within the boundaries of science.
7. I like to remind the frustrated researcher of Stern's (1965:773) sensitive admonition: "Whatever dangers personal weaknesses may present to the investigator, he can rise above them. He can retain the enthusiasm of youth which led him to contemplate the mysteries of the universe. He can remain grateful for the extraordinary privilege of participating in their exploration. He can incessantly find

delight in the discoveries made by other men, those of the past and those of his own times. And he can learn the difficult lesson that the journey itself and not only the great conquest is a fulfillment of human life."

8. See Hull (1988).

9. "To perceive a fact of nature which had never been seen before by any human eye or mind, to discover a new truth in any field, to uncover an event of past history or discern a hidden relation, such experiences the fortunate bearer will cherish throughout his life" (Stern 1965:772). Many scientists, in their autobiographies or in other writings, have extolled the joys of research (Shropshire 1981).

3. How Does Science Explain the Natural World?

1. Mayr (1964a, 1991) and Ghiselin (1969).

2. As stated by Kitcher (1993), philosophy of science "endeavored to analyze good science by focusing on questions of the confirmation of hypotheses by evidence, the nature of scientific laws and scientific theories, and the features of scientific explanation."

3. This work is not the place to present a history of the philosophy of science. The literature of this field is enormous, and I was not trained as a philosopher. My treatment is intended, rather, to reflect the viewpoint of the working scientist.

4. See Ghiselin (1969).

5. See Laudan (1968).

6. Among all the methods claimed to contribute to verification, the one which I trust the least is analogy. I am suspicious whenever someone tries to win an argument with the help of an analogy. Indeed, analogies are almost invariably misleading: they fail to be isomorphic with the real situation. Analogies are sometimes a useful didactic tool, permitting us to explain something unusual by comparing it with a familiar situation. However, they can never be treated as decisive evidence in an argument.

7. A theory normally remains in power until a better theory displaces it. There are a few exceptional cases, however, where all previous theories have been refuted decisively but no one has been able to come up with a credible replacement theory. The map sense in homing birds is an example of a problem that is without a current theory.

8. See Van Fraassen (1980).

9. Other semanticists have stressed that theories are formalized in set theory (whatever that means) instead of by axiomatization in mathematical logic, as in the received view. They employ "models"—"non-linguistic entities that are highly abstract and far removed from the empirical phenomena to which they will be applied" (Thompson 1989:69). Theories define a class of models; laws specify the behavior of a system. The problem with this terminology is that the set theory concept of a model is alien to most working biologists. For instance, I do not recall having encountered the term model even once in the entire classical evolutionary literature.

10. Fortunately for the nonphilosopher there are some excellent treatments of the history of these explanatory endeavors (for example, Suppe 1974; Kitcher and Salmon, eds., 1989).

11. This narrow focus of philosophy on justification to the exclusion of discovery has been criticized by Peirce (1972), Hanson (1958), Kuhn (1970), Feyerabend (1962, 1975), Kitcher (1993), and other philosophers.

12. Laudan (1977:198–225) gives an excellent analysis of this conflict. He rightly states that "until the rational history of any episode has been written, the cognitive sociologist must simply bite his tongue." "The chief reason for the sociologists' failure to find a correlation between scientific belief and social class is that the vast majority of scientific beliefs (though by no means all) seem to be of no social significance whatsoever."

13. See Mayr (1982:4).

14. See Junker (1995).

15. To take another example, for an extreme egalitarian, the idea of genetic differences among humans seems to be utterly distasteful. Laudan (1977) observes, "It has been suggested that any scientific theory that would argue for differences in ability or intelligence between the various [human] races must necessarily be unsound because such a doctrine runs counter to our egalitarian social and political framework."

16. I do not feel qualified to discuss other recent endeavors of explanation. However, it seems to me that the causal approaches of Laudan (1977), Salmon (1984, 1989), and Kitcher (1993) come perhaps closest to the actual practices of the working biologist. What was increasingly appreciated was that the assessment of a theory was not a matter of simple logical rules and that rationality had to be construed in broader terms than either deductive or inductive logic offered.

17. This was well recognized by Laudan (1977:3) when he said, "The rationality and progressiveness [I would simply say 'goodness'] of a theory are most closely linked—not with its confirmation or its falsification—but rather with its problem-solving effectiveness."

18. The question of the nature of realism, which has preoccupied philosophy to such an extent, has been remarkably irrelevant in the practical work of the scientists, and particularly so in that of the biologist. There is an enormous literature on realism. Some recent books are Harré (1986), Leplin (1984), McMullin (1988), Papineau (1987), Popper (1983), Putnam (1987), Rescher (1987), and Trigg (1989).

19. This situation is well understood by certain philosophers, for instance Hempel (1952) and Kagan (1989), while other philosophers have completely ignored the importance of precise, well-defined terminologies and of the avoidance of equivocation.

20. A similarly confusing change was the proposal of W. Hennig (1950) to transfer the term "monophyletic" from its traditional meaning as an attribute of a taxon to a new meaning as a process of descent. The confusion caused by this transfer can be avoided by using Ashlock's term holophyletic for Hennig's new concept (see Chapter 7).

21. Ghiselin (1984) has perceptively called attention to the frequency of such equivocations. It is a curious phenomenon that philosophers who pride themselves on the precision of their logic are anything but precise in their use of language. This has been justly castigated by the philosopher L. Laudan: "Philosophical dialogue is a curious activity. Arguments are expected to be rigorous, but no demand is made that there must be evidence for the premises. Terminology is expected to be precise, but its appropriateness to the subject matter under discussion can be left unexplored . . . and above all, the evidential warrant for one's philosophical claim is, like the topics of sex and religion to the less enlightened, one of those delicate issues never to be discussed in mixed company" (PSA 1978, vol. 2., 1979).

22. See Mayr (1986a, 1991, 1992b). Other examples are development (ontogeny vs. phylogeny), population (biological vs. mathematical set), species (typological vs. biological), function (physiological vs. ecological role), and gradualness (taxic vs. phenotypic).

23. Variety was used in zoology for a geographical race and hence potentially an incipient species, but this term was also used, particularly by botanists, for aberrant individuals within a population.

24. Much clarity was achieved in the taxonomic literature when the term taxon was adopted around 1950 for botanical and zoological groups and the term category restricted to rank in the Linnaean hierarchy, while previously the term category had been used for both. Recently Toulmin correctly calls attention to the fact that any word (term) employed in a theory carries some of its pre-theory meaning with it. This is particularly true when supporters and opponents of a theory hold very different Weltanschauungen. This is well illustrated by many biological controversies. For any teleologist—and most of Darwin's contemporaries were teleologists—selection meant something entirely different from Darwin's *a posteriori* description of differential survival and reproductive success. Species for an essentialist is something without essential variability and constant over time. It can change only by a saltation and is therefore incompatible with the biological species concept. One could tabulate all the terms involved in major scientific controversies and probably show that most of them had several meanings or connotations, depending on the Weltanschauung of the respective contestant.

25. There should never be any tension between the definition and the current scientific interpretation of the phenomenon to which the term is applied. The basic function of a definition is that of a heuristic device. Indeed, problems have sometimes been discovered when it was found that a traditional definition no longer fitted the subject matter. "Redefinitions in science are not complete breaks with the traditional definition but rather more precise formulations of terms that previously had been used vaguely or equivocally" (Ghiselin *in litt.*). Redefinitions are made possible by deeper analysis or new discoveries. For instance, Owen defined homology in terms of "the same" organ, without defining "the same," but Darwin's theory of common descent permitted a more precise definition. The redefinition should never involve replacement of the old by an entirely new concept.

26. As stated by Hempel (1952), "A real definition, according to traditional logic, is not a stipulation determining the meaning of some expression but a statement of the 'essential nature' or the 'essential attributes' of some entity." For the philosopher "definitions describe forms, and since forms are perfect and unchanging, definitions . . . are precise and rigorously-certain truths" (*Encyclopedia of Philosophy*).

27. Popper's confusion is well illustrated by his statement: "Never let yourself be goaded into taking seriously problems about words and their meanings. What must be taken seriously are questions of fact, and assertions about facts: theories and hypotheses; the problems they solve; and the problems they raise." This statement conceals the fact that in every theory and concept one must use words which one must define. One cannot argue about theories and hypotheses until one first clarifies what these theories are, and what the facts are. And since we use words to describe these theories and facts, we have to define them carefully, or else we risk equivocation. My previous examples (speciation, teleological, selection, etc.) have clearly shown how totally indispensable the need is for a clear definition of any word we use in a theory or explanation.

   Later in this chapter Popper sets meanings and truth in opposition to each other. He claims that the study of meanings leads to nothing and that in science everything has to do with the approach toward truth; and he emphasizes "in matters of the intellect, the only things worth striving for are true theories, or theories which come near to the truth." But he fails to see that one cannot have a true theory, let us say of speciation, if one has not previously established the meaning of the word speciation. Does one mean multiplication of species, or does one simply mean evolutionary change? Hence, it is quite evident that searching for meanings and searching for the truth are not two alternatives, but in fact the truth cannot be reached until we have clearly established the meaning of the words that are used. It is rather ironic that almost at the end of the chapter, which he calls "a long digression on essentialism," Popper remarks quite casually "we must understand the words in order to understand the theory." By that one single sentence he virtually takes back all the preceding claims of a strict alternative between meaning and truth. What Popper really says is exactly what I say, that we cannot establish the truth without having first established the meaning of the words we use. Ghiselin has pointed out very lucidly that one can give definitions only of concepts, but that actual particulars can only be described. Hence, one can define the species category but species taxa can only be named, described, and delimited.

28. One final note on language: When scientists are locked in a controversy over a particular issue, sometimes they choose words with negative connotations and apply them to the work of their opponents: "My work is dynamic; yours is static." "Mine is analytical; yours merely descriptive." "My explanation is mechanistic (that is, based on physical or chemical principles), while your explanation is holistic (that is, metaphysical)." The opponent usually has little trouble in reciprocating appropriately, but such exchanges of empty words rarely advance the long-term interests of science.

4. How Does Biology Explain the Living World?

1. Goudge (1961), Hull (1975b), Bock (1977), Nitecki and Nitecki (1992), and others.
2. See White (1965).
3. If speciation is a slow, gradual process, and if there are (as indeed, there are!) hundreds of thousands, if not millions, of populations (incipient species) in various stages of speciation at the present time, then it should be possible to reconstruct the entire process of speciation by placing the "stills" of various stages in the appropriate sequence. This is the same methodology that was used in the 1870s and 1880s by the students of cytology to reconstruct the process of cell division. They arranged hundreds of microscopic slides in a progressive sequence that would tell the story. I (Mayr 1942) did the same with natural populations representing all stages of "becoming species," and this has since been also done by scores of other authors (see also Mayr and Diamond 1997).
4. This leads us to the extremely complex philosophical problem of cause and causation. This book is not the place for a detailed analysis of this thorny problem. I will, therefore, not discuss Hume's critique of causation, according to which all we can determine is merely a sequence of events. I agree with those modern philosophers who admit that an antecedent event may have an effect, hence become a cause. Strictly causal sequences can be demonstrated particularly often in animal behavior. I find therefore nothing unscientific in the acceptance of commonsense causality.
5. Not that such case studies have not been provided previously, the outstanding example being the application of the semantic approach to evolutionary biology by Lloyd (1987). However, I will now present a number of cases of theory formation beginning with very simple ones and progressing to more complex cases. This will enable a philosopher who favors a particular approach to theory formation to test to what extent his approach is applicable to the particular case.
6. See also Mayr (1982, 1989a).
7. Lorenz's suggestion in its major points was adopted by Donald Campbell, Riedl, Oeser, Vollmer, Wuketis, Mohr, and many other biologists and philosophers.
8. See Kagan (1994).
9. There has been a curious controversy in the literature as to whether the human brain is adapted for the understanding of the mesokosmos. Those who denied this apparently had a teleological concept of selection and adaptation. But Darwinian adaptation is not teleological. There is no need to consider those individuals who survive the process of nonrandom elimination to be the products of a goal-directed process. An individual who survives the process of selection is adapted, one might say, by definition. The Darwinian is fully aware of the fact that all survivors owe their fate to a considerable extent also to stochastic processes. Accepting such a nonteleological concept of adaptation permits us to conclude: "Yes, the human brain is adapted for the understanding of the mesokosmos." All individuals that were inferior in this capacity were sooner or later eliminated without leaving descendants.

10. See Regal (1977).
11. Hamilton (1964).

5. Does Science Advance?

1. See Stent (1969).
2. The gradual advances in knowledge and understanding have been excellently described in a number of historical treatments. This includes the books by Hughes (1959), Baker (1948–1955), and Cremer (1985), as well as monographs by Coleman (1965), Churchill (1979), and others. For references, see Cremer volume.
3. Cremer (1985) describes in great detail these contributions.
4. Mayr (1982:810–811).
5. Technicians included Fol, Buetschli, Strasburger, van Beneden, and Flemming; theoreticians included Roux (1883), Weismann (1889), and Boveri (1903).
6. Hoyningen-Huene (1993) has presented an excellent analysis of Kuhn's views, including various changes after 1962. For an early set of criticisms see Lakatos and Musgrave (1970).
7. See Mayr (1991).
8. See Mayr (1972).
9. See Maynard Smith (1984:11–24).
10. Hoyningen-Huene (1993:197–206).
11. See Bowler (1983).
12. See Mayr (1946).
13. See Mayr (1990).
14. See Barrett et al. (1987).
15. This has been argued particularly strongly by Thagard (1992).
16. See Mayr (1952).
17. See Mayr (1992c).
18. See Mayr (1942).
19. This is the major subject of Hull's magisterial volume *Science as a Process* (1988).
20. Mayr (1954, 1963, 1982, 1989), Eldredge and Gould (1972), Stanley (1979).
21. What is accessible to science, and what is not, has been analyzed by Medawar (1984) and Rescher (1984). While many people such as DuBois-Reymond have underestimated the potential of science, many others tend to overestimate it.

6. How Are the Life Sciences Structured?

1. The traditional separation of zoology from botany survived in textbooks, teaching curricula, and library classifications long after it had been largely replaced by other ways of classifying the domain of biology. I know of only one work in the literature that is specifically devoted to a discussion of the structure of biology (Tschulok 1910), but it still accepts the traditional division of biology into botany and zoology, and is thus of little interest to a modern reader.

The terms zoology and botany, however, changed their meaning with the progress of biological research. Haeckel's *Generelle Morphologie* (1866) was remarkably Newtonian by defining nature as a system of forces inherent in matter. Consequently zoology had to be divided into morphology (the zoology of matter) and physiology (the zoology of forces). Under physiology Haeckel described also the relation of organisms to one another and to their environment, that is, ecology and biogeography. Ontogeny and phylogeny were included under morphology. The study of behavior was apparently ignored. Haeckel thus considered ecology, biogeography, and systematics as legitimate branches of biology, while the botanist Schleiden, in his reductionist attempt to reform botany, had no room in his system for the organismic aspects of plants (Schleiden 1842).

2. See Müller (1983).
3. Schleiden (1838) and Schwann (1839).
4. See Gerard (1958).
5. Weiss (1953:727).
6. The feasibility of reducing biology to physics was invariably illustrated by some simple physiological process, while completely ignoring evolutionary biology and other aspects of biology that cannot be reduced to physics (Nagel 1961). This attitude is well illustrated, for instance, by Needham, who in 1925 (244) described the recent changes in biology as a "change from comparative morphology to comparative biochemistry" and predicted that comparative biochemistry would eventually become transformed into electronic biophysics. He suggested that the interest in evolution should be replaced by the mechanistic theory of life. And since "mechanism is a more inclusive conception than evolution, it is deeper, and therefore it more definitely commands the cooperation of philosophy."
7. Handler (1970).
8. Lorenz (1973a) has rightly emphasized this point. Mainx (1955:3) has given a good account of the role of description in biological research.
9. Hennig (1950), Simpson (1961), Ghiselin (1969), Mayr (1969), Bock (1977), Mayr and Ashlock (1991), and Hull (1988).
10. See Mayr (1961).
11. Allen (1975:10).
12. "Structuralism assumes that there is a logical order to the biological realm and that organisms are generated according to rational dynamic principles" (Goodwin 1990).
13. "Only if [no explanations that can be deduced from general principles] can be found is an historical one accepted, faute de mieux" (Goodwin 1990:228).
14. The story of the respective contributions of botanists and zoologists to the advancement of biology is quite fascinating but has not yet been written. There was no real zoology prior to the nineteenth century, only its forerunners natural history and physiology (including embryology). Botany was clearly dominant owing to the prominent figure of Linnaeus. But the replacement of the downward classification of Linnaeus by upward classification was apparently mostly the work of zoologists in spite of the pioneering publications of Adanson and Jussieu.

Cytology was in an exemplary fashion the joint achievement of botanists (Schleiden) and zoologists (Schwann), with other botanists (e.g., Brown) and other zoologists (Meyen, Remak, Virchow) making significant contributions. Genetics is another field equally advanced by botanists (Mendel, DeVries, Johannsen, East, Correns, Müntzing, Nilsson-Ehle, Renner, Baur) and zoologists (Weismann, Bateson, Castle, Morgan, Chetverikov, Muller, Sonneborn), to mention only a few of the founders of this field.

15. This indispensability of the classical fields has been stressed by Stern (1962) and Mayr (1963a).

7. "What?" Questions: The Study of Biodiversity

1. This was followed by a period of intense preoccupation with phylogeny construction and macrotaxonomy, but basic taxonomy was rather neglected, if not despised in the heyday of experimental biology. The 1920–1950s experienced the flowering of the new systematics (Mayr 1942), followed in the 1960–1990s by the rise of numerical taxonomy and cladistics.

2. Simpson (1961).

3. See Mayr (1982:247–250) for a more detailed discussion of the contributions of taxonomy to the founding of new disciplines of biology.

4. That systematics is a field rich in theory was unfortunately ignored even by many systematists. The well-known ant specialist Wheeler stated in 1929, "Taxonomy . . . is the one biological science that has no theory, being merely diagnostics and classification" (1929:192).

5. See Mayr (1996).

6. How this inference is to be conducted is explained in the textbooks of taxonomy (Mayr and Ashlock 1991:100–105). Analogous difficulties are encountered by the paleontologist in the time dimension.

7. Sloan (1986).

8. Rosen (1979).

9. Mayr (1988a), Coyne et al. (1988).

10. According to Simpson, "Monophyly is the derivation of a taxon through one or more lineages from one immediately ancestral taxon of the same or lower rank" (1961:124). This definition articulates the traditional concept of monophyly that had been in use since Haeckel (1866). Cladists have transferred the term to a mode of descent (all taxa derived from an original stem species) but in order to prevent confusion with the traditional concept of monophyly, this cladistic concept should be called holophyly (Ashlock 1971).

11. Curiously, otherwise clearly homologous features are sometimes derived from different germ layers (see Chapter 8). Derivation from a given germ layer, therefore, is not necessarily a reliable indication of homology. Homology is always inferred.

12. The treatises of Simpson (1961), Mayr (1969), Bock (1977), and Mayr and Ashlock (1991) simply elaborate on Darwin's original two-criteria classification scheme.

13. Similarity is determined on the traditional criterion of taxonomists articulated by
    Whewell (1840:1:521) as follows: "The Maxim by which all Systems professing to
    be natural must be tested is this:—that the *arrangement obtained from one set of
    characters coincides with the arrangement obtained from another set*" (his italics).
    More or less the same idea was articulated by Hempel (1952:53), ". . . in so-called
    natural classifications the determining characteristics are associated, universally or
    in a high percentage of all cases, with other characteristics, of which they are
    logically independent." In contrast to cladistics, the traditional classification
    obeys Darwin's demand that "the different degrees of modification which [the
    diverging branches of the phylogenetic tree] have undergone . . . is expressed by
    the forms being ranked under different genera, families, sections or orders"
    (1859:420).
14. A higher category is defined as a class into which are placed all the higher taxa
    that are ranked at the same level in a hierarchic classification. The category species,
    for instance, is defined by the species definition, nowadays most often by the
    biological species definition.
15. The lack of correlation between evolutionary divergence and rate of speciation is
    also responsible for the so-called "hollow curve" (Mayr 1969).
16. Mayr (1995).
17. For instance, the clade recognized by the cladists that leads from the Pelycosauria
    to the mammals is primarily based on a single lower lateral temporal fenestra.
    Any single-character classification, even when strictly complying with phylogeny,
    results in artificial, heterogeneous taxa. Of course, the clade will acquire additional
    characters in due time, etc. etc.
18. Mayr (1995b).
19. Mayr and Bock (1994).
20. Mayr (1982:239–243), Mayr and Ashlock (1991:151–156).
21. A detailed explanation of the rules of zoological nomenclature is given by Mayr
    and Ashlock (1991:383–406).
22. Some authors recognize a third group, the Eocytes. Some specialists of the bacteria
    have claimed that the differences between the Archaebacteria and the Eubacteria
    are as great as that between the Prokaryota and the Eukaryota. There is no merit
    to that claim. The characterization of the bacteria in any classical textbook of
    microbiology applies equally well to both subdivisions of the Prokaryota, even
    though the Archaebacteria had not yet been characterized at that time. The
    Archaebacteria, no matter how different they are from the Eubacteria, even
    considering that the branching point of the two groups of Prokaryota is earlier
    than that between Prokaryota and Eukaryota, share most of their characters with
    the Eubacteria and should not be given the same high taxonomic rank as
    the Eukaryota. Renaming the Archaebacteria as Archaea cannot conceal the
    fact that like the Eubacteria, they are one of the two or three branches of the
    Bacteria.
23. For further details see Cavalier-Smith (1995a, 1995b) and Corliss (1994).

8. "How?" Questions: The Making of a New Individual

1. See Needham (1959) for an excellent presentation of Aristotle's ideas.
2. Today, we would say that these program-directed processes are teleonomic, but not teleological.
3. The foundations of the new descriptions of vertebrate development were laid by Pander (1817) but greatly improved and expanded by von Baer (1828ff).
4. Yolk-rich eggs often have a rather different development from yolk-poor eggs, even within a higher taxon. The total pathway of development is particularly different in organisms with different larval stages or complete metamorphosis. In the Lepidoptera, for instance, and other insect groups with complete metamorphosis, there is a total reorganization during the pupal stage and a new development of the adult structures from so-called imaginal discs.
5. This *vis essentialis* was, of course, a metaphysical *deus ex machina*, and Haller, a preformationist, was quite justified in asking, "Why should the unformed material coming from a hen always give rise to a chicken, and that from a peafowl give rise to a peafowl? To these questions no answer is given."
6. Moore (1993:445-456) provides an excellent summary of these researches.
7. Soon a connection between ontogeny and phylogeny was suggested with the gastrula stage corresponding to the coelenterate type, and later stages of development representing the "types" of "higher" organisms. Haeckel, more than anyone else, emphasized such a recapitulationary aspect of development and proposed the gastraea theory of the evolution of the invertebrates.
8. Saha (1991:106).
9. Not only are genes composite, consisting of exons that are transcribed and introns that are excised prior to protein synthesis, but in addition to the enzyme-producing structural genes there are regulatory genes and flanking sequences. All this is far too complex a machinery to be described in detail in this volume, and I must refer to the appropriate literature such as *The Molecular Biology of the Gene* (Alberts et al. 1983).
10. This has been stressed particularly by Severtsov and his school (Schmalhausen).
11. How well they knew the fact that embryos did not correspond to the adult stages of the ancestors is clearly stated in the writings of Haeckel and others.
12. Mayr (1954).
13. For more detailed studies of development, I recommend Davidson (1986), Edelman (1988), Gilbert (1991), Hall (1992), Horder et al. (1986), McKinney et al. (1991), Moore (1993), Needham (1959), Russell (1916), Slack et al. (1993), and Walbot et al. (1987).

9. "Why?" Questions: The Evolution of Organisms

1. The origin of life is a chemical process, involving autocatalysis and some direction-giving factor. As Eigen showed, it would seem that prebiotic selection must

have been involved no matter which particular pathway of origin of life is postulated. For details, see Shapiro (1986) and Eigen (1992).

2. Alfred Russel Wallace proposed that isolating mechanisms were produced by natural selection, but Darwin vigorously opposed this idea. Right to the present day, there have been two camps, consisting of the Wallace and the Darwin followers, on this question. Dobzhansky followed Wallace, while H. J. Muller and Mayr followed Darwin.

3. See Alexander (1987), Trivers (1985), Wilson (1975).

4. This approach was taken by Rensch (1939, 1943) and Simpson (1944), who showed that macroevolutionary phenomena could be considered as consistent with the findings of genetics. In particular, it was possible to explain all so-called evolutionary laws, such as Cope's law or Dollo's law, in terms of variation and selection.

5. Mayr (1954:206–207). See also p. 172, above.

10. What Questions Does Ecology Ask?

1. The heterogeneity of the subject matter bracketed under the name of ecology has long been realized. This is why there are now separate texts for evolutionary ecology, behavioral ecology, population biology, limnology, marine ecology, and paleoecology. To add to this diversity there are vast differences in the ecology of different groups of animals, plants, and microorganisms, and of different environmental realms. Terrestrial ecology is very different from freshwater ecology (limnology) and marine ecology. Plankton ecology, founded by V. Hensen, became a flourishing science of great importance for fisheries. Anyone who wants to be a well-rounded ecologist must familiarize himself with an enormous range of subject matter. This diversity is part of the reason for the numerous difficulties encountered in the study of ecology, as will be discussed in the next sections. There are numerous investigations of the ecological or natural history knowledge of certain periods, as those of Cittadino (1990) and Egerton (1968, 1975).

And it is true for this area, as it is sometimes said, that everything interacts with everything else. The whole, now called ecology, "is held together more by the adoption of a name and the cohesion of professional societies than by a commonality of philosophy or purpose. Thus, ecology poses special difficulties for the historian" (Ricklefs 1985:799). There are a number of rather simple definitions of ecology, such as "the relations of organisms to their environment," but this permits an enormous range of possible inclusion. Every structure of an organism, each of its physiological properties, all of its behavior, and indeed, almost any component of its phenotype and genotype has evolved for an optimal relation of the organism to its environment.

As a result, there are vast areas of overlap between ecology and other biological disciplines such as evolutionary biology, genetics, behavior, and physiology. For instance, in Ricklefs's (1990) comprehensive ecology text, six entire chapters are devoted to evolutionary questions, chapters that could with equal justification be part of a text on evolutionary biology. A number of recently published texts have been simply called Evolutionary Ecology. They deal with such topics as extinction, adaptation, life histories, sex, social behavior, and coevolution. All physiological adaptations of organisms for their specialized mode of life or for the specialized environments in which they live are rightly considered by Ricklefs to be the concern of ecology. So are all adaptations permitting organisms to cope with extreme climatic conditions, such as daily and seasonal cycles, migrations, and other behavioral adaptations. There are numerous physiological mechanisms in the service of environmental adaptations, particularly for extreme environments, such as deserts or the Arctic (Schmidt-Nielsen 1990). Adaptation to local conditions is well illustrated in plants by the development of ecotypes.

2. Glacken (1967) has given us a detailed documentation of man's concepts of the environment from antiquity to the end of the eighteenth century. There are numerous investigations of the ecological or natural history knowledge of certain periods, as those of Egerton (1968, 1975).

3. See Stresemann (1975).

4. In 1949 the multiauthored *Principles of Animal Ecology* of the Chicago school (AEPPS) was published and from that date on a stream of new textbooks of ecology has appeared; no less than six such texts were reviewed in a single issue of *Science* (Orians 1973). Outstanding among these was Eugene Odum's *Fundamentals of Ecology*, published in 1953 and most widely adopted until the 1970s, and Robert Ricklefs' *Ecology* (1973), perhaps now the most widely used text in the United States. The growth of the field is illuminated by the fact that Odum's first edition had 384 pages while the third edition of Ricklefs (1990) has 896 pages. It is evident that only a fraction of the aspects and problems of ecology can be dealt with in the present short overview.

5. The rebellion against the purely descriptive approach of systematics and morphology, indicated by the flourishing of experimental researches in physiology and embryology (Entwicklungsmechanik), was matched in natural history by an emphasis on the relations of whole living organisms. Anything having to do with the living organism was referred to in Germany as *Biologie*, in a meaning quite different from the term biology for the combination of zoology and botany that was traditional in the English-language literature. The volume on animal life (by Doflein) in the famous Hesse-Doflein set, which was a splendid summary of the prevailing knowledge of the living animals and plants, was strongly influenced by Darwinian thinking. This *Biologie* saw itself as an alternative to and supplemental to morphology, the study of "dead structures." Its subject matter was more or less that which in modern textbooks is treated under the headings of behavioral and evolutionary ecology. This biology dealt almost exclusively with animals.

6. See Kingsland (1985).

7. Historically, population biology was long considered an independent branch of biology, but it is now clear that it is a branch of ecology, as was particularly emphasized at the Cold Spring Harbor Symposium of 1957.

8. V. C. Wynne-Edwards (1962, 1986).

9. What remained of the close relation between taxonomy and ecology has been dealt with in a number of publications, for instance, by Heywood (1973).

10. Sometimes ecologists apply the term "population" to multispecies assemblages in an ecosystem. He may speak of the plankton population of a lake or the herbivore population of a savanna. In most cases such use of the term population for portions of a multispecies ecosystem is misleading.

11. There was a minor development of a similar interest in animals reflected in the publication of R. Hesse's *Tiergeographie auf Ökologischer Grundlage* (1924). In spite of its title, this was not an animal geography dealing with the distribution of animals and the causes of their distribution, but rather it was an animal ecology as affected by geographical factors. In some respects it was a successor to Semper's (1881) ecological morphology. In due time community ecology gave rise to ecosystem ecology (see below).

12. "The climax formation is the adult organism, the fully developed community."

13. See Mayr (1941), MacArthur and Wilson (1963), and Mayr (1965).

11. Where Do Humans Fit into Evolution?

1. These last two species are sometimes assigned to a separate genus, *Paranthropus*.

2. With *robustus* restricted to south Africa and *boisei* to east Africa, it is impossible to say which is morphologically more similar to their common ancestor, although the greater age of *A. aethiopicus* suggests that in many ways *A. robustus* is more derived.

3. The evidence for the recency of the branching of the hominid line from the chimpanzee line has steadily improved. It began with work on blood proteins (Goodman), followed by DNA hybridization tests by Sibley and Ahlquist, later confirmed by Caccone and Powell (with improved methods), and finally by other molecular methods and chromosomes.

4. Sarich (1967) was the first to make this claim. Additional fossil finds are necessary to pinpoint this date more narrowly.

5. In order to try to straighten out the then-existing chaos of hominid classification (over 30 generic and over 100 specific names), I applied Occam's razor and proposed in 1950 that only a single hominid species lived at any one time in the past, just as there is at present only one species of *Homo*. Subsequent researches have shown that my proposal was a considerable oversimplification.

6. See Mayr (1954).

7. See Stanley (1992).

8. Donald (1991).

9. See Mayr (1963:650).

10. Mitton (1977).

11. Mayr (1982:623–624).

12. Haldane (1949).

12. Can Evolution Account for Ethics?

1. "All that we know . . . [shows] that from the remotest times successful tribes have supplanted other tribes" (1871:160).

2. Altruism in social animals does not necessarily involve a disadvantage for the altruist. Darwin stated this very nicely, "We have now seen that actions are regarded by savages, and were probably so regarded by primeval man, as good or bad, solely as they affect in an obvious manner the welfare of the tribe" (1871:96). Darwin expressed the close relation between sociality and ethical norms by claiming "that the so-called moral sense is aboriginally derived from the social instincts" (1871:97).

3. De Waal (1996).

4. Wilson (1993) has given us an excellent presentation of the evidence for the existence of a moral sense in mankind. See Bradie (1994).

5. Sulloway (1996).

6. Kohlberg (1981; 1984).

7. The topic of evolution and ethics has produced an enormous literature during the last 20 years, to a considerable extent stimulated by E. O.Wilson's *Sociobiology* (1975). Authors who, in addition to Wilson, have made major contributions to the subject are R. D. Alexander, A. Gewirth, R. J. Richards, M. Ruse, and G. C. Williams. Presentations of their views with bibliographies of their writings, as well as a number of classical essays (T. H. Huxley, J. Dewey), and 10 essays by other authors are published in the volume *Evolutionary Ethics* by Nitecki and Nitecki (1993). This is a very useful introduction to the literature of evolutionary ethics.

Bibliography

Adanson, M. 1763. *Familles des Plantes*. Paris.

Agar, W. E. 1948. "The wholeness of the living organism." *Phil. Sci.* 15:179–191.

Alberts, B., D. Bray, J. Lewis, K. Roberts, and J. Watson. 1983. *Molecular Biology of the Cell.* 1st ed. New York and London: Garland.

Alexander, R. D. 1987. *The Biology of Moral Systems.* Hawthorne, N.Y.: Aldine de Gruyter.

Allee, W. C., A. E. Emerson, O. Park, T. Park, and K. P. Schmidt. 1949. *Principles of Animal Ecology.* Philadelphia: Saunders.

Allen, G. E. 1975. *Life Science in the Twentieth Century.* New York: John Wiley & Sons.

Alvarez, L. 1980. "Asteroid theory of extinctions strengthened." *Science* 210:514.

Ashlock, P. 1971. "Monophyly and associated terms." *Syst. Zool.* 21:430–438.

Avery, O. T., C. M. MacLeod, and M. McCarty. 1944. "Studies on the chemical nature of the substance inducing transformation of pneumococcal types." *J. Exp. Med.* 79:137–158.

Ayala, F. J. 1987. "The biological roots of morality." *Biol. and Phil.* 2:235–252.

Ayala F. J., A. Escalante, C. O'Huigin, and J. Klein. 1994. "Molecular genetics of speciation and human origins." *Proc. Nat. Ac. Sci.* 91:6787–6794.

Baer, K. E. von. 1828. *Entwicklungsgeschichte der Thiere: Beobachtung und Reflexion.* Königsberg: Bornträger.

Baker, J. R. 1938. "The evolution of breeding searson." In G. R. de Beer, ed., *Evolution: Essays on Aspects of Evolutionary Biology,* pp. 161–177. Oxford: Clarendon Press.

——— 1948–1955. "The cell theory: a restatement, history, and critique." *Quart. J. Microscopical Science* 89:103–123; 90:87–108; 93:157–190; 96:449.

Barrett, P. H., P. J. Gautrey, S. Herbert, D. Kohn, and S. Smith. 1987. *Charles Darwin's Notebooks, 1836–1844.* Ithaca: Cornell University Press.

Bates, H. W. 1862. "Contributions to an insect fauna of the Amazon Valley." *Trans. Linn. Soc. London* 23:495–566.

Bateson, P., ed. 1983. *Mate Choice.* Cambridge: Cambridge University Press.

Beatty, J. 1995. "The evolutionary contingency thesis." In G. Wolters and J. Lennox, eds., *Concepts, Theories, and Rationality in the Biological Sciences*, pp. 45–81. Pittsburgh: University of Pittsburgh Press.

Beckner, M. 1959. *The Biological Way of Thought*. New York: Columbia University Press.

———— 1967. "Organismic biology." In *Encyclopedia of Philosophy*, vol. 5., pp. 549–551.

Bertalanffy, L. von. 1952. *Problems of Life*. London: Watts.

Blandino, G. 1969. *Theories on the Nature of Life*. New York: Philosophical Library.

Blumenbach, J. F. 1790. *Beyträge zur Naturgeschichte*. Göttingen.

Bock, W. 1977. "Foundations and methods of evolutionary classification." In M. Hecht, P. C. Goody, and B. M. Hecht, eds., *Major Patterns in Vertebrate Evolution*, pp. 851–895. New York: Plenum Press.

Bowler, P. J. 1983. *The Eclipse of Darwinism: Anti-Darwinian Evolution Theories in the Decades around 1900*. Baltimore: Johns Hopkins University Press.

Boveri, T. 1903. "Über den Einflus der Samenzelle auf die Larvencharaktere der Echiniden." *Roux's Arch.* 16:356.

Bradie, M. 1994. *The Secret Chain*. Albany: State University of New York Press.

Buffon, G. L. 1749–1804. *Histoire naturelle, générale et particulière*. 44 vols. Paris: Imprimerie Royale, puis Plassan.

Carr, E. H. 1961. *What Is History?* London: Macmillan.

Cassirer, E. 1950. *The Problem of Knowledge: Philosophy, Science, and History since Hegel*. New Haven: Yale University Press.

Cavalier-Smith, T. 1995a. "Membrane heredity, symbiogenesis, and the multiple origins of algae." In Arai, Kato, and Dio, eds., *Biodiversity and Evolution*, pp. 69–107. Tokyo: The National Science Museum Foundation.

———— 1995b. "Evolutionary protistology comes of age: biodiversity and molecular cell biology." *Arch. Protistenkd* 145:145–154.

Cittadino, E. 1990. *Nature as the Laboratory*. New York: Columbia University Press.

Churchill, F. B. 1979. "Sex and the single organism: biological theories of sexuality in mid-nineteenth century." *Stud. Hist. Biol.* 3:139–177.

Code. 1985. *International Code of Zoological Nomenclature*. Adopted by the General Assembly of the International Union of Biological Sciences. Berkeley: University of California Press.

Coleman, W. 1965. "Cell nucleus and inheritance: an historical study." *Proc. Amer. Philos. Soc.* 109:124–158.

Coon, C. 1962. *The Origin of Races*. New York: Alfred A. Knopf.

Corliss, J. O. 1994. "An interim utilitarian ('user-friendly') hierarchical classification of the protista." *Acta Protozoologica* 33:1–51.

Coyne, J. A., H. A. Orr, and D. J. Futuyma. 1988. "Do we need a new definition of species?" *Syst. Zool.* 37:190–200.

Cremer, T. 1985. *Von der Zellenlehre zur Chromosomentheorie*. Berlin: Springer.

Crick, F. 1966. *Of Molecules and Men*. Seattle: University of Washington Press.

Darwin, C. 1859. *On the Origin of Species by Means of Natural Selection or the Preservation of Favored Races in the Struggle for Life*. London: Murray. Facsimile edition 1964, ed. E. Mayr.

———— 1871. *The Descent of Man*. London: Murray.

———— 1994. *The Correspondence of Charles Darwin,* vol. 9: 269 [letter to Henry Fawcett, 18 Sept. 1861]. Cambridge: Cambridge Univeresity Press.

Davidson, E. H. 1986. *Gene Activity in Early Development,* 3rd ed. Orlando: Academic Press.

De Waal, Franz. 1996. *Good Natured: The Origins of Right and Wrong in Humans and Other Animals.* Cambridge: Harvard University Press.

Diamond, J. 1991. *The Third Chimpanzee: The Evolution and Future of the Human Animal.* New York: HarperCollins.

Dijksterhuis, E. J. 1961. *The Mechanization of the World Picture,* trans. C. Dikshoorn. Oxford: Clarendon Press.

Dobzhansky, T. 1937. *Genetics and the Origin of Species.* New York: Columbia University Press.

———— 1968. "On Cartesian and Darwinian aspects of biology." *Graduate Journal* 8:99–117.

———— 1970. *Genetics of the Evolutionary Process.* New York: Columbia University Press.

Doflein, F. 1914. *Das Tier als Glied des Naturganzen.* Leipzig: Teubner.

Donald, Merlin. 1991. *Origins of the Modern Mind: Three Stages in the Evolution of Culture and Cognition.* Cambridge: Harvard University Press.

Driesch, H. 1905. *Der Vitalismus als Geschichte und als Lehre.* Leipzig: J. A. Barth.

———— 1908. *The Science and Philosophy of the Organism.* London: A. and C. Black.

DuBois-Reymond, E. 1860. "Gedächtnisrede auf-Johannes Müller." *Abt. Presa. Aked. Wiss.* 1859:25–191.

———— 1872. *Über die Grenzen des Naturwissenschaftlichen Erkennens.* Leipzig.

———— 1887. *Die Sieben Welträtsel.* Leipzig.

Dupré, J. 1993. *The Disorder of Things.* Cambridge: Harvard University Press.

Edelman, G. 1988. *Topobiology: An Introduction to Molecular Embryology.* New York: Basic Books.

Egerton, F. N. 1968. "Studies of animal populations from Lamarck to Darwin." *J. Hist. Biol.* 1:225–259.

———— 1975. "Aristotle's population biology." *Arethusa* 8:307–330.

Eigen, M. 1992. *Steps toward Life.* Oxford: Oxford University Press.

Eldredge, N. 1971. "The allopatric model and phylogeny in Paleozoic invertebrates." *Evolution* 25:156–167.

Eldredge, N., and S. J. Gould. 1972. "Punctuated equilibria: an alternative to phyletic gradualism," in Schopf 1972, pp. 82–115.

Elton, C. 1924. "Periodic fluctuations in the numbers of animals: their causes and effects." *J. Exper. Biol.* 2:119–163.

———— 1927. *Animal Ecology.* New York: Macmillan.

Evans, F. C. 1956. "Ecosystem as the basic unit in ecology." *Science* 123:1127–1128.

Feyerabend, P. 1962. "Explanation, reduction, and empiricism." *Minnesota Studies Philos. Sci.* 2:28–97.

———— 1970. "Against method: Outline of an anarchistic theory of knowledge." *Minnesota Studies Philos. Sci.* 4:17–130.

———— 1975. *Against Method.* London: Verso.

Frege, G. 1884. *Die Grundlagen der Arithmetik: Eine logisch mathematische Untersuchung über den Begriff der Zahl.* Breslau: W. Koebner.

Geoffroy St. Hilaire, E. 1818. *Philosophie anatomique.* Paris.

Gerard, R. W. 1958. "Concepts and principles of biology." *Behavioral Science* 3:95–102.

Ghiselin, M. T. 1969. *The Triumph of the Darwinian Method.* Berkeley: University of California Press.

——— 1974. *The Economy of Nature and the Evolution of Sex.* Berkeley: University of California Press.

——— 1984. "'Defnition,' 'character,' and other equivocal terms." *Syst. Zool.* 33:104–110.

——— 1989. "Individuality, history, and laws of nature in biology." In M. Ruse, ed., *What the Philosophy of Biology Is,* pp. 3–66. Dordrecht: Kluwer.

Giere, R. N. 1988. *Explaining Science: A Cognitive Approach.* Chicago: University of Chicago Press.

Gilbert, S., ed. 1991. *A Conceptual History of Modern Embryology.* New York: Plenum.

Glacken, C. J. 1967. *Traces on the Rhodian Shore: Nature and Culture in Western Thought.* Berkeley: University of California Press.

Gleason, H. A. 1926. "The individualistic concept of the plant association." *Bull. Torrey Bot. Club* 53:7–26.

Goldschmidt, R. 1938. *Physiological Genetics.* New York: McGraw-Hill.

——— 1954. "Different philosophies of genetics." *Science* 119:703–710.

Goodwin, B. 1990. "Structuralism in biology." *Sci. Progress* (Oxford) 74:227–244.

Goudge, T. A. 1961. *The Ascent of Life.* Toronto: University of Toronto Press.

Graham, L. R. 1981. *Between Science and Values.* New York: Columbia University Press.

Haeckel, E. 1866. *Generelle Morphologie der Organismen: Allgemeine Grundzüge der organischen Formen-Wissenschaft, mechanisch begründet durch die von Charles Darwin reformirte Descendenz-Theorie.* 2 vols. Berlin: Georg Reimer.

——— 1870 (1869). "Ueber Entwickelungsgang u. Aufgabe der Zoologie." *Jenaische Z.* 5:353–370.

Haldane, J. B. S. 1949. "Human evolution: past and future." In Jepsen, Mayr, and Simpson 1949:405–418.

Haldane, J. S. 1931. *The Philosophical Basis of Biology.* London: Hodder and Stoughton.

Hall, B. K. 1992. *Evolutionary Developmental Biology.* London: Chapman and Hall.

Hall, R. 1954. *The Scientific Revolution, 1500–1800.* London: Longmans.

Hall, T. S. 1969. *Ideas of Life and Matter.* 2 vols. Chicago: University of Chicago Press.

Hamilton, W. D. 1964. "The genetical evolution of social behavior." *J. Theoret. Biol.* 7:1–16; 17–52.

Handler, P., ed. 1970. *The Life Sciences.* Washington, D.C.: National Academy of Sciences.

Hanson, N. R. 1958. *Patterns of Discovery.* Cambridge: Cambridge University Press.

Haraway, D. J. 1976. *Crystals, Fabrics, and Fields.* New Haven: Yale University Press.

Harper, J. L. 1977. *Population Biology of Plants.* New York: Academic Press.

Harré, R. 1986. *Varieties of Realism: A Rationale for the Natural Sciences.* Oxford: Oxford University Press.

Hempel, C. G. 1952. *Fundamentals of Concept Formation in Empirical Science.* Chicago: University of Chicago Press.

———— 1965. *Aspects of Scientific Explanation.* New York: Free Press.

Hempel, C. G., and P. Oppenheim. 1948. "Studies in the logic of explanation." *Phil. Sci.* 15:135–175.

Hennig, W. 1950. *Grundzüge einer Theorie der Phylogenetischen Systematik.* Berlin: Deutscher Zentralverlag.

Hertwig, O. 1876. "Beiträge zur Kenntnis der Bildung, Befruchtung und Theilung des thierischen Eies." *Morph. Jahrb.* 1:347–434.

Hesse, R. 1924. *Tiergeographie auf Ökologischer Grundlage.* Jena: Fischer.

Heywood, V. H. 1973. *Taxonomy and Ecology: Proceedings of an International Symposium Held at the Dept. of Botany, University of Reading.* New York: Systematics Association by Academic Press.

Holton, G. 1973. *Thematic Origins of Scientific Thought: Kepler to Einstein.* Cambridge: Harvard University Press.

Horder, T. J., H. A. Witkowski, and C. C. Wylie, eds. 1986. *A History of Embryology.* New York: Cambridge University Press.

Hoyningen-Huene, P. 1993. *Reconstructing Scientific Revolutions: Thomas S. Kuhn's Philosophy of Science.* Chicago: University of Chicago Press.

Hughes, A. 1959. *A History of Cytology.* London and New York: Abelard-Schuman.

Hull, D. L. 1975. "Central subjects and historical narratives." *History and Theory* 14:253–274.

———— 1988. *Science as a Process: An Evolutionary Account of the Social and Conceptual Development of Science.* Chicago: University of Chicago Press.

Humboldt, A. von. 1805. *Essay sur la Geograpahie des Plantes.* Paris.

Huxley, J. S. 1942. *Evolution, the Modern Synthesis.* London: Allen & Unwin.

Huxley, T. H. 1863. *Evidence as to Man's Place in Nature.* London: William and Norgate.

———— 1893. *Evolution and Ethics.* Romanes Lecture. London: Oxford University Press.

Jacob, François. 1973. *The Logic of Life: A History of Heredity.* New York: Pantheon.

———— 1977. "Evolution and tinkering." *Science* 196:1161–1166.

Jepsen, G. L., E. Mayr, and G. G. Simpson. 1949. *Genetics, Paleontology, and Evolution.* Princeton University Press.

Johannsen, W. 1909. *Elemente der Exakten Erblichkeitslehre.* Jena: Gustav Fischer.

Junker, Thomas. 1995. "Darwinismus, materialismus und die revolution von 1848 in Deutschland. Zur interaktion von politik und wissenschaft." *Hist. Phil. Life Sci.* 17:271–302.

Kagan, J. 1989. *Unstable Ideas.* Cambridge, Mass.: Harvard University Press.

———— 1994. *Galen's Prophesy: Temperament in Human Nature.* New York: Basic Books.

Kant, I. 1790. *Kritik der Urteilskraft.* Berlin.

Kimura, M. 1983. *The Neutral Theory of Molecular Evolution.* Cambridge: Cambridge University Press.

Kingsland, S. E. 1985. *Modeling Nature: Episodes in the History of Population Ecology.* Chicago: University of Chicago Press.

Kitcher, P. 1993. *The Advancement of Science.* New York: Oxford University Press.

Kitcher, P., and W. L. Salmon, eds. 1989. *Scientific Explanation*. Minneapolis: University of Minnesota Press.

Kohlberg, L. 1981. *The Philosophy of Moral Development: Moral Stages and the Idea of Justice*. New York: Harper & Row.

―――― 1984. *The Psychology of Moral Development: The Nature and Validity of Moral Stages*. San Francisco: Harper & Row.

Kölliker, A. von. 1841. *Beiträge zur Kenntniss der Geschlechtsverhältnisse und der Samenflüssigkeit wirbelloser Thiere, nebst einem Versuch über das Wesen und die Bedeutung der sogenannnten Samenthiere*. Berlin: W. Logier.

―――― 1886. "Das Karyoplasma und die Vererbung." In *Kritik der Weismann'schen Theorie von der Kontinuitat des Keimplasma*. Leipzig.

Kölreuter, J. G. 1760. See Mayr 1986a.

Korschelt, E. 1922. *Lebensdauer Altern und Tod*. Jena: Gustav Fisscher.

Kuhn, T. 1962. *The Structure of Scientific Revolutions*. Chicago: University of Chicago Press.

―――― 1970. *Reflections on my Critics*. In Lakatos and Musgrave 1970, pp. 231–278.

La Mettrie, J. O. de. 1748. *L'homme machine*. Leyden: Elie Luzac.

Lack, D. 1954. *The Natural Regulation of Animal Numbers*. Oxford: Clarendon Press.

Lakatos, I., and A. Musgrave, eds. 1970. *Criticism and the Growth of Knowledge*. Cambridge: Cambridge University Press.

Lamarck, J. B. 1809. *Philosophie zoologique, ou exposition der considérations relatives à l'histoire naturelle des animaux*. Paris.

Laudan, L. 1968. "Theories of scientific method from Plato to Mach." *Hist. Sci.* 7:1–63.

―――― 1977. *Progress and Its Problems: Towards a Theory of Scientific Growth*. Berkeley: University of California Press.

Lenoir, T. 1982. *The Strategy of Life*. Dordrecht: D. Reidel.

Leplin, J., ed. 1984. *Scientific Realism*. Berkeley: University of Califorinia Press.

Liebig, J. 1863. *Ueber Francis Bacon von Verulam und die Methode von Naturforschung*. Munich: J. G. Cotta.

Lindeman, R. L. 1942. "The trophic-dynamic aspect of ecology." *Ecology* 23:399–418.

Lorenz, K. 1973. "The fashionable fallacy of dispensing with description." *Naturwiss.* 60:1–9.

Lloyd, E. 1987. *The Structure of Evolutionary Theory*. Westport, Conn.: Greenwood Press.

Lyell, C. 1830–1833. *Principles of Geology, Being an Attempt to Explain the Former Changes of the Earth's Surface, by Reference to Causes Now in Operation*. 3 vols. London.

MacArthur, R. H., and E. O. Wilson. 1963. "An equilibrium theory of insular zoogeography." *Evolution* 17:373–387.

Magnol, P. 1689. *Prodromus historiae generalis plantarum in quo familiae plantarum per tabulas disponuntur*. Montpellier.

Maier, A. 1938. *Die Mechanisierung des Weltbildes. Forschungen zur Geschichte der Philosophie und der Pädagogik*. Leipzig.

Mainx, F. 1955. "Foundations of biology." *Int. Encycl. Unif. Sci.* 1:1–86.

May, R. M. 1973. *Stability and Complexity in Model Ecosystems*. Princeton: Princeton University Press.

Maynard Smith, J., Jr. 1984. "Science and myth." *Natural History* 11:11–24.

Mayr, E. 1941. "The origin and the history of the bird fauna of Polynesia." *Proc. Sixth Pacific Sci, Congress.* 4:197–216.

———— 1942. *Systematics and the Origin of Species.* New York: Columbia University Press.

———— 1946. "History of the North American bird fauna." *The Wilson Bulletin* 58:3–41.

———— 1952. "The problem of land connections across the South Atlantic, with special reference to the Mesozoic." *Bulletin of the American Museum of Natural History* 99:85, 255–258.

———— 1954. "Change of genetic environment and evolution." In J. Huxley, A. C. Hardy, and E. B. Ford, eds., *Evolution as a Process.* London: Allen & Unwin, pp. 157–180.

———— 1961. "Cause and effect in biology: kinds of causes, predictability, and teleology are viewed by a practicing biologist." *Science* 134:1501–1506.

———— 1963a. *Animal Species and Evolution.* Cambridge: The Belknap Press of Harvard University Press.

———— 1963b. "The new versus the classical in science." *Science* 141, no. 3583:765.

———— 1964. "Introduction." In C. Darwin, *On the Origin of Species: A Facsimile of the First Edition,* pp. vii–xxv. Cambridge: Harvard University Press.

———— 1965. "Avifauna: turnover on islands." *Science* 150:1587–1588.

———— 1969. *Principles of Systematic Zoology.* New York: McGraw-Hill.

———— 1972. "The nature of the Darwinian revolution: acceptance of evolution by natural selection required the rejection of many previously held concepts." *Science* 176:981–989.

———— 1976. *Evolution and the Diversity of Life:* Selected Essays. Cambridge: The Belknap Press of Harvard University Press.

———— 1982. *The Growth of Biological Thought: Diversity, Evolution, and Inheritance.* Cambridge: The Belknap Press of Harvard University Press.

———— 1986a. "Joseph Gottlieb Kölreuter's contributions to biology." *Osiris* 2d ser. 2:135–176.

———— 1986b. "Natural selection: the philosopher and the biologist." Review of Sober. *Paleobiology* 12:233–239.

———— 1988. "The why and how of species." *Biol. and Phil.* 3:431–441.

———— 1989. "Speciational evolution or punctuated equilibria." *Journal of Social and Biological Structures* 12:137–158.

———— 1990. "Plattentektonik und die Geschichte der Vogelfaunen." In R. van den Elzen, K.-L. Schuchmann, and K. Schmidt-Koenig, eds., *Current Topics in Avian Biology,* pp. 1–17. Proceedings of the International Centennial Meeting of the Deutsche Ornithologen-Gesellschaft, Bonn 1988. Bonn: Verlag der Deutschen Ornithologen-Gesellschaft.

———— 1991a. *One Long Argument: Charles Darwin and the Genesis of Modern Evolutionary Thought.* Cambridge: Harvard University Press.

———— 1991b. "The ideological resistance to Darwin's theory of natural selection." *Proceedings of the American Philosophical Society* 135:123–139.

———— 1992a. "The idea of teleology." *Journal of the History of Ideas* 53:117–135.

—— 1992b. Darwin's principle of divergence. *Journal of the History of Biology* 25:343–359.

—— 1995a. "Darwin's impact on modern thought." *Proceedings of the American Philosophical Society* 139(4):317–325. (Read 10 November, 1994.)

—— 1995b. "Systems of ordering data." *Biol. and Phil.*: 10(4):419–434.

—— 1996. "What is a species and what is not?" *Phil. of Sci.* 63(2):261–276.

Mayr, E., and P. Ashlock. 1991. *Principles of Systematic Zoology*, rev. ed. New York: McGraw-Hill.

Mayr, E., and W. Bock. 1994. "Provisional classifications v standard avian sequences: heuristics and communication in ornithology." *Ibis* 136:12–18.

Mayr, E., and J. Diamond. 1997. *The Birds of Northern Melanesia*. Oxford: Oxford University Press.

McKinney, M. L., and K. J. McNamara. 1991. *Heterochrony: The Evolution of Ontogeny*. New York: Plenum.

McLaughlin, P. 1991. "Newtonian biology and Kant's mechanistic concept causality." In G. Funke, ed., *Akten Siebenten Internationalen Kant Kongress*, pp. 57–66. Bonn: Bouvier.

McMullin, E., ed. 1988. *Construction and Constraint: The Shaping of Scientific Rationality*. Notre Dame, Ind.: Notre Dame University Press.

Medawar, P. B. 1984. *The Limits of Science*. Oxford: Oxford University Press.

Mendel, J. G. 1866. "Versuche über Pflanzen-hybriden." *Verh. Natur. Vereins Brünn* 4(1865):3–57.

Merriam, C. H. 1894. "Laws of temperature control of the geographic distribution of terrestrial animals and plants." *Nat. Geogr. Mag.* 6:229–238.

Meyen, F. J. F. 1837–1839. *Neues System der Pflanzenphysiologie*. 3 vols. Berlin: Haude und Spenersche Buchhandlung.

Michener, C. D. 1977. "Discordant evolution and the classification of allodapine bees." *Syst. Zool.* 26:32–56; 27:112–118.

Milkman, R. D. 1961. "The genetic basis of natural variation III." *Genetics* 46:25–38.

Miller, S. J. 1953. "A production of amino acids under possible primitive earth conditions." *Science* 117:528.

Mitton, J. B. 1977. "Genetic differentiation of races of man as judged by single-locus or multiple-locus analyses." *Amer. Nat.* 111:203–212.

Moore, J. A. 1993. *Science as a Way of Knowing*. Cambridge: Harvard University Press.

Morgan, C. L. 1923. *Emergent Evolution*. London: William and Norgate.

Müller, G. H. 1983. "First use of *biologie.*" *Nature* 302:744.

Munson, R. 1975. "Is biology a provincial science?" *Phil. Sci.* 42:428–447.

Nagel, E. 1961. *The Structure of Science: Problems in the Logic of Scientific Explanation*. New York: Harcourt, Brace & World.

Nägeli, C. W. 1845. "Über die gegenwärtige Aufgabe der Naturgeschichte, insbesondere der Botanik." *Zeitschr. Wiss. Botanik*, vols. 1 and 2. Zürich.

—— 1884. *Mechanisch-physiologische Theorie der Abstammungslehre*. Leipzig: Oldenbourg.

Needham, J., ed. 1925. *Science, Religion and Reality*. London: The Sheldon Press.

—— 1959. *A History of Embryology*. 2nd ed. New York: Abelard-Schuman.

Nitecki, M. H., and D. V. Nitecki. 1992. *History and Evolution.* Albany: State University of New York Press.

——— 1993. *Evolutionary Ethics.* Albany: State University of New York Press.

Novikoff, A. 1945. "The concept of integrative levels and biology." *Science* 101:209–215.

Odum, E. P. 1953. *Fundamentals of Ecology.* Philadelphia: Saunders.

Orians, G. H. 1962. "Natural selection and ecological theory." *Amer. Nat.* 96:257–264.

Pander, H. C. 1817. *Beiträge zur Entwicklungsgeschichte des Hühnchens im Eye.* Würzburg.

Papineau, D. 1987. *Reality and Representation.* Oxford: Clarendon Press.

Pearson, K. 1892. *The Grammar of Science.* London: W. Scott.

Peirce, C. S. 1972. *The Essential Writings,* ed. E. C. Moore. New York: Harper & Row.

Polanyi, M. 1968. "Life's irreducible structure." *Science* 160:1308–1312.

Popper, K. 1952. *The Open Society and Its Enemies.* London: Routledge & Kegan Paul.

——— 1968. *Logic of Scientific Discovery.* New York: Harper & Row.

——— 1974. *Unended Quest: An Intellectual Autobiography.* La Salle, Ill.: Open Court.

——— 1975. *Objective Knowledge: An Evolutionary Approach.* Oxford: Clarendon Press.

——— 1983. *Realism and the Aim of Science.* New Jersey: Rowan & Littefield..

Putnam, H. 1987. *The Many Faces of Realism.* La Salle, Ill.: Open Court.

Redfield, R., ed. 1942. "Levels of integration in biological and social sciences." *Biological Symposia VIII.* Lancaster, Penn.: Jacques Cattell Press.

Regal, P. J. 1975. "The evolutionary origin of feathers." *Quarterly Review of Biology* 50:35–66.

——— 1977. "Ecology and evolution of flowering plant dominance." *Science* 196:622–629.

Remak, R. 1852. "Über extracellulare Entstehung thierischer Zellen und über Vermehrung derselben durch Theilung." *Archiv für Anatomie, Physiologie und wissenschaftliche Medicin (Müllers Archiv)* 19:47–72.

Rensch, B. 1939. "Typen der Artbildung." *Biol. Reviews* (Cambridge) 14:180–222.

——— 1943. "Die biologischen Beweismittel der Abstammungslehre." In G. Heberer, *Evolution der Organismen,* pp. 57–85. Jena: Gustav Fischer.

——— 1947. *Neuere Probleme der Abstammungslehre.* Stuttgart: Enke.

——— 1968. *Biophilosophie.* Stuttgart: Gustav Fischer.

Rescher, N. 1984. *The Limits of Science.* Berkeley: University of California Press.

——— 1987. *Scientific Realism: A Critical Reappraisal.* Dordrecht: Reidel.

Ricklefs, R. E. 1990. *Ecology,* 3rd ed. New York: Freeman (1st ed. 1973).

Ritter, W. E., and E. W. Bailey. 1928. "The organismal conception: its place in science and its bearing on philosophy." *Univ. Calif. Pub. Zool.* 31:307–358.

Rosen, D. 1979. "Fishes from the upland intermountain basins of Guatemala." *Bull. Amer. Mus. Nat. His.* 162:269–375.

Rosenfield, L. L. 1941. *From Beast-Machine to Man-Machine.* New York: Oxford University Press.

Roux, W. 1883. *Über die Bedeutung der Kerntheilungsfiguren.* Leipzig: Engelmann.

——— 1895. *Gesammelte Abhandlungen über Entwicklungsmechanik der Organismen.* 2 vols. Leipzig: Engelmann.

———— 1915. "Das Wesen des Lebens." *Kultur der Gegenwart* III 4(1):173–187.

Ruse, M. 1979a. *Sociobiology: Sense or Nonsense?* Boston: D. Reidel.

———— 1979b. *The Darwinian Revolution.* Chicago: University of Chicago Press.

Russell, E. S. 1916. *Form and Function: A Contribution to the History of Animal Morphology.* London: J. Murray.

———— 1945. *The Directiveness of Organic Activities.* Cambridge: Cambridge University Press.

Saha, M. 1991. "Spemann seen through a lens." In S. F. Gilbert, ed., *Developmental Biology: A Conceptual History of Modern Embryology,* pp. 91–108. New York: Plenum Press.

Salmon, W. C. 1984. *Scientific Explanation and the Causal Stuctures of the World.* Princeton: Princeton University Press.

———— 1989. *Four Decades of Scientific Explanation.* Minneapolis: University of Minnesota Press.

Sarich, V. M., and A. C. Wilson. 1967. "Immunological time scale for hominid evolution." *Science* 158:1200–1202.

Sattler, R. 1986. *Biophilosophy.* Berlin: Springer.

Schleiden, M. J. 1838. "Beiträge zur Phytogenesis." *Archiv für Anatomie, Physiologie und wissenschaftliche Medicin (Müllers Archiv)* 5:137–176.

———— 1842. *Grundzüüge der wissenschaftlichen Botanik.* Leipzig.

Schmidt-Nielsen, K. 1990. *Animal Physiology: Adaptation and Environment.* 4th ed. Cambridge: Cambridge University Press.

Schopf, Thomas J. M., ed. 1972. *Models in Paleobiology.* San Francisco: Freeman.

Schwann, Th. 1839. *Mikroskopische Untersuchungen über die Übereinstimmung in der Struktur und dem Wachstum der Tiere und Pflanzen.* Berlin.

Semper, K. G. 1881. *Animal Life as Affected by the Natural Conditions of Existence.* New York: Appleton [1880 in German].

Severtsoff, A. N. 1931. *Morphologische Gesetzmässigkeiten der Evolution.* Jena: Gustav Fischer.

Shapiro, J. H. 1986. *Origins: A Skeptic's Guide to the Creation of Life on Earth.* New York: Summit Books.

Shropshire, W., Jr. 1981. *The Joys of Research.* Washington, D.C.: Smithsonian Institution Press.

Simpson, G. G. 1944. *Tempo and Mode in Evolution.* New York: Columbia University Press.

———— 1961. *Principles of Animal Taxonomy.* New York: Columbia University Press.

———— 1969. "Biology and ethics." In G. G. Simpson, ed., *Biology and Man,* pp. 130–148. New York: Harcourt, Brace and World.

Singer, P. 1981. *The Expanding Circle.* New York: Farrar, Straus and Giroux.

Slack, J. M., P. W. Holland, and C. F. Graham. 1993. "The zootype and the phylotypic stage." *Nature* 361:490–492.

Sloan, P. R. 1986. "From logical universals to historical individuals: Buffon's idea of biological species." In J. Roger and J. L. Fischer, eds., *Histoire des concepts d'espèce dans la science de la vie.* Paris: Fondation Singer-Polignac.

Smart, J. J. C. 1963. *Philosophy and Scientific Realism.* London: Routledge & Kegan Paul.

Smuts, J. C. 1926. *Holism and Evolution.* New York: Viking Press. 2nd ed. 1965.

Snow, C. P. 1959. *The Two Cultures and the Scientific Revolution.* New York: Cambridge University Press.

Spemann, H. 1901. "Über Correlationen in der Entwicklung des Auges." *Verhandl Anat Ges.* 15:15–79.

Spemann, H., and H. Mangold. 1924. "Über Induktion von Embryoanlagen durch Implantation artfremder Organisatoren." *Roux's Archiv* 100:599–638.

Stanley, S. M. 1979. *Macroevolution: Pattern and Process.* San Francisco: W. H. Freeman.

——— 1992. "An ecological theory for the origin of Homo." *Paleobiology* 18:237–257.

Stebbins, G. L. 1950. *Variation and Evolution in Plants.* New York: Columbia University Press.

Stent, G. 1969. *The Coming of the Golden Age: A View of the End of Progress.* New York: Natural History Press.

Stern, C. 1962. "In praise of diversity." *Am. Zool.* 2:575–579.

——— 1965. "Thoughts on research." *Science* 148:772–773.

Stresemann, E. 1975. *Ornithology: From Aristotle to the Present.* Cambridge: Harvard University Press.

Sulloway, Frank. 1996. *Born to Rebel.* New York: Pantheon Press.

Suppé, F., ed. 1974. *The Structure of Scientific Theories.* Urbana: University of Illinois Press. 2nd ed. 1977.

Tansley, A. G. 1935. "The use and abuse of vegetational concepts and terms." *Ecology* 16:204–307.

Thagard, P. 1992. *Conceptual Revolutions.* Princeton: Princeton University Press.

Thompson, P. 1988. "Conceptual and logical aspects of the 'new' evolutionary epistemology." *Can. J. Phil.,* suppl vol. 14:235–253.

——— 1989. *The Structure of Biological Theories.* Albany: State University of New York Press.

Thoreau, H. D. 1993 [ca. 1856–1862]. *Faith in a Seed.* Washington, D.C.: Island Press.

Thornton, Ian. 1995. *Krakatau: The Destruction and Reassembly of an Island Ecosystem.* Cambridge: Harvard University Press.

Treviño, S. 1991. *Graincollection: Human's Natural Ecological Niche.* New York: Vintage Press.

Treviranus, G. R. 1802. *Biologie, oder Philosophie der lebenden Natur.* Vol. 1. Göttingen: J. R. Röwer.

Trigg, R. 1989. *Reality at Risk: A Defense of Realism in Philosophy and the Sciences.* 2nd ed. New York: Harvester Wheatsheaf.

Trivers, R. L. 1985. *Social Evolution.* Menlo Park: Benjamin/Cummings.

Tschulok, S. 1910. *Das System der Biologie in Forschung und Lehre.* Jena: Gustav Fischer.

Van Fraassen, B. C. 1980. *The Scientific Image.* Oxford: Clarendon Press.

Waddington, C. H. 1960. *The Ethical Animal.* London: Allen and Unwin.

Walbot, V., and N. Holder. 1987. *Developmental Biology.* New York: Random House.

Warming, J. E. B. 1896. *Lehrbuch der ökologischen Pflanzengeographie.* Berlin.

Weismann, A. 1883. *Über die Vererbung.* Jena: Gustav Fischer.

——— 1889. *Essays upon Heredity.* Oxford: Clarendon Press.

Weiss, P. 1947. "The place of physiology in the biological sciences." *Federation Proceedings* 6:523–525.

———— 1953. "Medicine and society: the biological foundations." *J. Mount Sinai Hospital* 19:727.

Wheeler, W. H. 1929. "Present tendencies in biological theory." *Sci. Monthly* 1929:192.

Whewell, W. 1840. *Philosophy of the Inductive Sciences Founded upon Their History.* Vol. 1. London: J. W. Parker.

White, M. 1965. *Foundations of Historical Knowledge.* New York: Harper and Row.

Wilson, E. B. 1925. *The Cell in Development and Heredity.* 3rd ed. New York: Macmillan.

Wilson, E. O. 1975. *Sociobiology.* Cambridge: Harvard University Press.

Wilson, J. Q. 1993. *The Moral Sense.* New York: Free Press.

Windelband, W. 1894. "Geschichte der alten Philosophie: Nebst einem Anhang: Abriss der Geschichte der Mathematik und der Naturwissenschaften." In *Altertum von Siegmund Günter.* 2 vols. Munich: Beck.

Wolff, C. F. 1774. *Theoria generationis.* Halle.

Woodger, J. H. 1929. *Biological Principles: A Critical Study.* London: Routledge and Kegan Paul.

Wynne-Edwards, V. C. 1962. *Animal Dispersion in Relation to Social Behavior.* Edinburgh: Oliver & Boyd.

———— 1986. *Evolution through Group Selection.* Oxford: Blackwell Scientific Press.

Glossary

Acquired characters Those characteristics of an organism's phenotype that result from environmental influences rather than inheritance

Adaptationist program The research endeavor to discover the adaptive significance of structures, processes, and activities

Adaptedness The suitability of a structure or an organism for its environment or lifestyle, as a result of past selection

Adaptive zone A resource space in the environment occupied by organisms more or less specially adapted for it

Allopatric speciation Geographic speciation

Altruism Behavior that benefits another organism, at some cost to the actor

Animism The belief that phenomena in nature are inhabited by spirits

Apomorphy A derived state in an evolutionary series of homologous characters

Artificial selection Selection of breeding stock by an animal or plant breeder

Asexual reproduction Any form of propagation not resulting from the fusion of two gametes

Autecology The ecology of species (and of individuals)

Autogenetic theories Theories based on a belief in goal-directed forces or tendencies in living nature

Autotrophs Organisms capable of self-producing their nutritive needs, as plants with the help of sunlight

Base pair A pair of hydrogen-bonded nitrogenous bases (one purine and one pyrimidine) that connect the two strands of the DNA double helix

Batesian mimicry Imitation by a palatable species of the appearance of a nonpalatable or toxic species

Bauplan Structural type, as that of a vertebrate or arthropod

Biological species concept Definition of a species as a reproductively (genetically) isolated group of interbreeding natural populations

Biota Fauna and flora

Blending inheritance The now-discredited notion that the paternal and maternal genetic materials fuse during fertilization; see Particulate inheritance

Cartesianism The beliefs, methods, and philosophy of Descartes

Catastrophism The theory that catastrophic events in the history of the earth have resulted in the partial or complete extinction of the biota

Category In taxonomy, the rank (such as species, genus, family, order) assigned to a taxon in the Linnaean hierarchy

Central dogma The now-proven assertion that the information contained in proteins cannot be translated back into nucleic acids

Character A component of the phenotype

Chromatin The material of which chromosomes are composed, including DNA and proteins

Chromosome One of the threadlike structures in the nucleus of the cell, consisting of DNA and associated proteins

Cladification A system of ordering organisms in which the to-be-ordered items are branches of the phylogenetic tree (or of a cladogram); Hennigian classification

Cladistic analysis Analysis of the derived characters of organisms to infer the branching sequence in phylogeny based exclusively on derived characters

Cladogram Inferred branching pattern of a phylogenetic tree

Cladon Taxon based on the principles of Hennigian cladification

Classification (Darwinian) An ordering of species or higher taxa into groups (classes) on the basis of both similarity (degree of evolutionary divergence) and common descent (genealogy)

Common descent The derivation of species or higher taxa from a common ancestor

Competitive exclusion The principle that no two species with identical ecological requirements can coexist at the same place; also called Gause's principle

Convergence In evolution, the independent acquisition of the same feature by two or more unrelated lineages

Creationism Belief in the literal truth of the story of creation as recorded in the Book of Genesis

Cytoplasm Contents of the cell around the nucleus

Deme A local population of a species; the community of potentially interbreeding individuals at a given locality

Determinate development Development in which the fate of embryonic cells is determined by their position in the developing embryo, and each region of the embryo differentiates almost independently from any influence of other regions; also called mosaic development

Determinism A theory that the outcome of any process is strictly predetermined by definite causes and natural laws and is therefore, in theory, predictable

Dichopatric speciation Speciation achieved through the division of a parental species by a geographical, vegetational, or other extrinsic barrier

Diploid Having a doubled set of chromosomes, one from each parent

DNA Deoxyribonucleic acid, the molecule transmitting genetic information

DNA hybridization Method to test the closeness of relationship of two taxa

Ectoderm The outer germ layer, usually giving rise to the epidermis and the neural system

Ectotherm An organism whose temperature is determined by the temperature of its environment

Emergence In systems, the occurrence of characters at higher levels of integration which could not have been predicted from a knowledge of lower-level components

Endoderm The inner germ layer, usually giving rise to the intestinal system

Epigenesis The now-discredited theory that new structures originate during ontogeny from undifferentiated material with the help of a vital force; see Preformation

Epistatic interactions Interaction of different gene loci

Essentialism A belief that the variation of nature can be reduced to a limited number of basic classes, representing constant, sharply delimited types; typological thinking

Eukaryotes Organisms with a well-developed nucleus; all organisms above the level of prokaryotes

Evolutionary causation The historical factors responsible for the properties of individuals and species, and more specifically for the composition of the genotype (the genetic program)

Evolutionary synthesis The period between 1937 and 1950 when conceptual unity among evolutionists was established, based on an essentially Darwinian paradigm including natural selection, adaptation, and the study of diversity

Exon A sequence of base pairs in a gene that participates in the coding of proteins (peptides); see Intron

Female choice The hypothesis that it is often the female that selects one of several available males for mating, rather than the other way around; it is part of the modern theory of sexual selection

Finalism Belief in an inherent trend in the natural world toward some preordained final goal or purpose, such as the attainment of perfection; see Teleology

Fitness The relative ability of an organism to survive and transmit its genes to the gene pool of the next generation

Founder population A population founded by a single female (or a small number of conspecifics) beyond the previous species border

Functional causation Proximate causation

Gamete A germ cell (egg or sperm) carrying half of the organism's full set of chromosomes, especially a mature germ cell capable of participating in fertilization; see Genetic recombination; Meiosis

Gene A sequence of base pairs in a DNA molecule that contains information for the construction of one protein molecule

Genetic code The code by which the genetic information contained in the base-pair sequence of DNA is translated into amino acids (the building blocks of proteins)

Genetic drift Changes in the gene content of a population owing to chance events

Genetic program The information coded in an organism's DNA

Genetic recombination The reshuffling of an organism's genes during meiosis; it ensures that the chromosomes carried by an organism's eggs or sperm are not identical to the chromosomes the organism inherited from either of its own parents, and that no two chromosomes of any of the eggs or sperm are likely to be identical

Genome The totality of genes carried by a single gamete

Genotype The totality of the genes (genetic information) of an individual

Geographic speciation Speciation that occurs while populations are geographically isolated; also known as allopatric speciation

Germ cell An egg or sperm cell

Germ plasm An outmoded term referring to the genetic material in the germ cells

Gradualism A theory that evolution progresses by the gradual modification of populations, and not by the sudden origin of new types (saltations)

Guild A group of species with similar resource requirements and foraging methods and therefore having similar roles in the ecosystem and being potential competitors of one another

Haploid Having a single set of chromosomes

Holophyletic taxa All the taxa descended from a single stem species

Homologous characters Features in two or more taxa that had descended from the same feature of the nearest common ancestor

Homoplasy Possession by two or more taxa of a character not derived from the nearest common ancestor but acquired through convergence, parallelism, or reversal

Imprinting A particularly rapid and largely irreversible learning process that stores information in an open program

Incipient species A population in the process of evolving into a separate species

Inclusive fitness Additions to the fitness of the genotype of an individual that are provided by the genotypes of close relatives, particularly descendants

Inheritance of acquired characters The now-discredited theory that changes in an organism's phenotype caused by factors in its environment can be passed on to offspring through the organism's genetic material

Intron A noncoding sequence of base pairs that is eliminated prior to the translation of the nucleic acids into proteins (peptides); see Exon

Isolating mechanisms Genetic (including behavioral) properties of individuals which prevent populations of different species that coexist in the same area from interbreeding

Kin selection In individuals related by common descent, selection for the shared components of their genotypes

Lamarckism Lamarck's evolutionary theories, particularly the belief in the inheritance of acquired characters

Macroevolution Evolution above the species level, the evolution of higher taxa, and the production of evolutionary novelties, such as new structures

Macrotaxonomy The classification of higher taxa

Meckel-Serrès law Recapitulation

Meiosis Two consecutive special cell divisions in the developing germ cells, characterized by the pairing and segregation of homologous chromosomes; the resulting germ cells have a haploid set of chromosomes

Mesoderm The middle germ layer, giving rise to connective tissue, muscles, bones

Metazoan A multicellular animal

Mesozoic The geological era that lasted from about 225 million years to 65 million years; the age of reptiles

Microtaxonomy Classification at the species level

Mitosis The process by which a cell (including its chromosome set) divides into two daughter cells

Monophyly The descent of a taxon from the nearest common ancestral taxon of the same or lower rank

Morphotype The structural type or bauplan

Mosaic development Determinate development

Mutation A spontaneous or induced change in the DNA sequence of a gene in an individual organism; ordinarily mutations are the result of an error in DNA replication

Natural kinds Types of organisms, as defined by the typological species concept

Natural selection The nonrandom survival and reproductive success of a small percentage of the individuals of a population owing to their possession, at that moment, of characters that enhance their ability to survive and reproduce

Natural theology The study of nature to document evidence for the power and wisdom of the Creator in the design of the world

Neutral evolution The occurrence and accumulation of heritable mutations which do not change the fitness of the individual or its offspring

Niche The multidimensional resource space of a species; its ecological requirements

Ontogeny The development of the individual from the fertilized egg (zygote) to adulthood

Open program A set of tissues able to incorporate and retain instructions with which to influence development and activities

Organicism The belief that the unique characteristics of living organisms are due not to their composition but rather to their organization

Orthogenesis The belief in an intrinsic force or tendency that carries a phylogenetic lineage toward a predetermined goal or at least to greater perfection

Pangenesis A theory attempting to explain the inheritance of acquired characters by proposing that small granules (gemmules, pangenes) migrate from all parts of the body to the gonads, where they are incorporated in the gametes

Parapatric speciation The progressive divergence into two separate species of two populations having contiguous geographic ranges but no (or only minimal) interbreeding in the zone of contact

Paraphyletic A taxon that includes a lineage leading to a derived taxon

Parsimony In taxonomy, the principle that the shortest tree is the best, that is, the tree that has the smallest number of branching points (character changes)

Parthenogenesis The production of offspring from unfertilized eggs

Particulate inheritance The now-proven theory that the genetic materials contributed by the parents do not fuse during fertilization but remain discrete; see Blending inheritance

Peripatric speciation The origin of new species through the modification of peripherally isolated founder populations (budding)

Phenetics The delimitation and ranking of taxa based strictly on overall similarity without evaluation of genealogy

Phenotype The totality of the characteristics of an individual, resulting from the interaction of the genotype with the environment

Phylogeny The pathway of descent from ancestors

Physicalism An emphasis on (and belief in) certain principles dominant in classical physics, such as essentialism, determinism, reductionism, and so on

Pleiotropic Of a gene affecting several phenotypic characters

Plesiomorphic An ancestral (primitive, patristic) character state

Polygenic characters Aspects of the phenotype controlled by several genes

Polyploidy The possession of more than two haploid chromosome sets

Polytypic species Species composed of several subspecies

Population thinking A viewpoint that emphasizes the uniqueness of every individual in populations of a sexually reproducing species and therefore the real variability of populations; the opposite of essentialism and typological thinking

Preformation The now-discredited theory that an embryo develops from material in which the essential form of the adult is "preformed," that is, already exists in its essential structures; see Epigenesis

Primate Member of the order of mammals that includes lemurs, monkeys, and apes

Prokaryotes One-celled organisms lacking a structured nucleus, such as various kinds of bacteria

Protists A heterogeneous assemblage of one-celled eukaryotes

Proximate causation Chemical and physical factors responsible for biological processes, that is, for activities resulting from the decoding of the genetic program

Punctuationism The theory that most evolutionarily important events take place during short bouts of speciation, and that once species are formed they are relatively stable, sometimes for very long periods

Recapitulation The theory that organisms recapitulate during their ontogeny the phylogenetic stages through which their ancestors had passed; also known as the Meckel-Serrès law

Reduction division A cell division in meiosis during which the number of chromosomes is halved, that is, a diploid cell gives rise to two haploid cells

Reductionism A philosophy which states that all phenomena and laws relating to complex phenomena (including living ones) can be explained only by reducing them to their smallest components and that the higher levels of integration of these systems can be fully explained through a knowledge of the smallest components

Regulative development Development in the early embryo in which the cellular environment influences each cell

Reversal The reappearance in phylogeny of an ancestral character as a result of the loss of a derived (apomorphic) character

Saltationism The belief that evolutionary change is the result of the sudden origin of a new kind of individual which becomes the progenitor of a new kind of organism

Scala naturae A linear arrangement of the forms of life from the lowest, nearly inanimate, to the most perfect; the great chain of being

Sexual selection Selection for characters enhancing reproductive success

Sibling species Reproductively isolated but morphologically identical or nearly identical species

Sister groups Groups originating through the split of a phylogenetic lineage

Sociobiology The systematic study of the biological basis of social behavior, with special emphasis on reproductive behavior

Soft inheritance The now-discredited notion that acquired characters of the phenotype can be transmitted to the genotype; see Central dogma

Somatic program In development, the information contained in adjacent tissues that may influence or control the further development of an embryonic structure or tissue

Speciational evolution Rapid evolution of species that had originated as peripatric founder populations

Species (biological) A reproductively isolated aggregate of populations which can interbreed with one another because they share the same isolating mechanisms

Species concept The biological meaning or definition of the word "species"

Species category The category in the Linnaean hierarchy in which species taxa are ranked

Species taxa Particular populations or groups of populations that comply with the species definition

Stasis The maintenance of a constant phenotype by an evolutionary lineage through geological time

Stem species A species with a new apomorphy giving rise to a new clade

Stochastic processes Chance events

Stratigraphy The study of geological strata, their history, and the fossil fauna and flora contained in them

Sympatric speciation Speciation without geographical isolation, perhaps by ecological specialization; the acquisition of isolating mechanisms within a deme

Synecology The ecology of communities and ecosystems

Systematics The science which studies the diversity of organisms

Taxon A monophyletic group of organisms that share a definite set of characters and are sufficiently distinct to be worthy of a formal name

Taxonomy The theory and practice of classifying organisms

Teleology The actual or only seeming existence of end-directed processes in nature, and their study

Teleonomic process A process or behavior that owes its goal-directedness to the operation of a program

Tertiary The most recent of the major geological eras, extending from about 65 million years ago to between 500,000 and 2,000,000 years ago

Transposons Genes that move from one chromosome to another

Typological thinking Essentialism

Typological species concept Definition of species on the basis of degree of difference

Ultimate causations Evolutionary causations

Uniformitarianism A theory, promoted particularly by the geologist Charles Lyell, that all changes in nature are gradual, particularly geological ones; the opposite of catastrophism

Variant A member of a variable population

Vicariance The existence of closely related forms (vicariants) in unconnected geographical areas which had been secondarily isolated by the formation of a natural barrier

Vitalism The belief that living organisms have a special vital force or vital substance that cannot be found in inert matter

Zygote A fertilized egg; the individual that results from the union of two gametes and their nuclei

Guide to Topics Covered

Acknowledgments

In the completion of this wide-ranging book, I have received a great deal of help and encouragement from many colleagues. Walter J. Bock read the entire manuscript of an earlier and a later version and made numerous valuable recommendations. Several chapters benefited significantly from critical reviews by David Pilbeam and Richard Alexander. The area where I needed most guidance was in the philosophy of science. It is not easy for a working scientist to follow the line of argument of the various, often drastically disagreeing, schools of epistemology. Professors Adolf Grünbaum and John Beatty have patiently explained matters to me that I had not before been able to understand. I have also received valuable advice from David Hull, Michael Ruse, and Robert Brandon. I regret that some of their suggestions could not be incorporated, owing to limitations of space. To all of these friends I am deeply indebted.

Much of the writing was done during my winter trips to the South. I wish to express by warm appreciation to Dr. Ira Rubinoff, Director of the Smithsonian Tropical Research Institute (STRI) in Panama, for accepting me for several years as a guest researcher. Dr. John Fitzpatrick and his staff also have my sincere thanks for being my hosts at the Archbold Biological Station at Lake Placid, Florida. Professors Karl Peters and Dan DeNicola provided me with working facilities at the Department of Philosophy and Religion at Rollins College, Winter Park, Florida; to them and to Rollins College President Rita Bornstein, who appointed me Johnston Visiting Scholar for 1995, I wish to say thank you again for the generosity I have been shown.

My devoted late secretary, Walter Borawski, was as helpful as ever in the earlier phases of preparing the manuscript. I was extremely fortunate in having Lisa Reed as his efficient and hardworking successor; she not only typed numerous versions of the manuscript and compiled much of the bibliography but also prepared careful subject indexes that helped eliminate overlap among the chapters. Chenoweth Moffatt was responsible for typing the final version, completing the bibliography, and making the manuscript ready for the publisher. To all of these assistants I owe a great debt of gratitude for their intelligent and faithful support.

The staff of Harvard University Press once again has made every effort to produce a book of the highest quality, for which I am very grateful. Most of all, I wish to thank Susan Wallace Boehmer, my editor since 1982, whose good counsel throughout the past year has substantially improved the organization and readability of this book.

It is my sincere hope that *This Is Biology* will contribute to a better understanding not only of biology but of science as a whole. Such an understanding is necessary if we are to develop the values by which this country and the entire world will have to live in the future.

Cambridge, Massachusetts
September 1996

Index